Drilling and Excavation Technologies for the Future

Committee on Advanced Drilling Technologies

Geotechnical Board/Commission on Engineering and Technical Systems

Board on Earth Sciences and Resources/Commission on Geosciences, Environment, and Resources

National Research Council

NATIONAL ACADEMY PRESS
WASHINGTON, D.C. 1994

NOTICE: The project that is the subject of this report was approved by the Governing Board of the National Research Council, whose members are drawn from the councils of the National Academy of Sciences, the National Academy of Engineering, and the Institute of Medicine. The members of the panel responsible for this report were chosen for their special expertise and with regard for appropriate balance among government, industry, and academia.

This report has been reviewed by a group other than the authors according to procedures approved by a Report Review Committee consisting of members of the National Academy of Sciences, the National Academy of Engineering, and the Institute of Medicine.

Support for this study was provided by the Department of Energy (grant number DE-FG42-92R208008) and the Gas Research Institute. Any opinions, findings, conclusions, or recommendations expressed herein are those of the committee and do not necessarily reflect the views of the Department of Energy or the Gas Research Institute.

Library of Congress Catalog Card No. 94-66775
International Standard Book Number 0-309-05076-6

Copies of this report are available from:
National Academy Press
2101 Constitution Avenue, N.W.
Box 285
Washington, DC 20055
800-624-6242 or 202-334-3313 (in the Washington metropolitan area)

B-330

Cover art by Shelley Myers, Project Assistant for the Board on Earth Sciences and Resources, National Research Council. Myers' work is exhibited widely in the Washington, D.C. area and has won several area awards.

Printed in the United States of America

COMMITTEE ON ADVANCED DRILLING TECHNOLOGIES

Ali S. Argon, Massachusetts Institute of Technology
(*Chair since July 1993*)
Neville G.W. Cook, University of California, Berkeley
(*Chair until July 1993*)
George A. Cooper, University of California, Berkeley
Michael M. Herron, Schlumberger-Doll Research
Stephen E. Laubach, The University of Texas, Austin
William C. Maurer, Maurer Engineering, Inc.
James E. Monsees, Parsons Brinckerhoff, Inc.
D. Stephen Pye, UNOCAL Corporation
Jean-Claude Roegiers, University of Oklahoma
Eugene D. Shchukin, Institute of Physical Chemistry RAS
Mark D. Zoback, Stanford University

NRC Staff
Peter H. Smeallie, Study Director
Thomas M. Usselman, Senior Program Officer
Kevin D. Crowley, Program Officer
Jennifer T. Estep, Administrative Assistant
Judith L. Estep, Administrative Assistant
Amelia B. Mathis, Senior Secretary
Nathan L. Harshman, Research Aide

LIAISONS

The U.S. Department of Energy

Stephen Brocoum, Office of Civilian Radioactive Waste Management, Washington, D.C.

William J. Gwilliam, Morgantown Energy Technology Center, Morgantown, West Virginia

Allan J. Jelacic, Office of Energy Efficiency and Renewable Energy, Washington, D.C.

Roy C. Long, Yucca Mountain Project Office, Las Vegas, Nevada

William C. Luth, Office of Energy Research, Washington, D.C.

John E. Mock, Office of Energy Efficiency and Renewable Energy, Washington, D.C.

Albert B. Yost II, Morgantown Energy Technology Center, Morgantown, West Virginia

Gas Research Institute

Kent F. Perry, Gas Research Institute, Chicago, Illinois

Massachusetts Institute of Technology

Carl Peterson, Department of Mechanical Engineering

Jeff Tester, Energy Laboratory

GEOTECHNICAL BOARD

COMMISSION ON ENGINEERING AND TECHNICAL SYSTEMS

BOARD ON EARTH SCIENCES AND RESOURCES

J. Freeman Gilbert, ***Chair*****,** University of California, San Diego
Gail M. Ashley, Rutgers University
Thure Cerling, University of Utah
Mark P. Cloos, University of Texas, Austin
Neville G.W. Cook, University of California, Berkeley
Joel Darmstadter, Resources for the Future
Donald J. DePaolo, University of California, Berkeley
Marco T. Einaudi, Stanford University
Norman H. Foster, Independent Geologist
Charles G. Groat, Louisiana State University
Donald C. Haney, Kentucky Geological Survey
Andrew H. Knoll, Harvard University
Philip E. LaMoreaux, P.E. LaMoreaux and Associates, Inc.
Susan Landon, Thomasson Partner Associates
Marcia K. McNutt, Massachusetts Institute of Technology
J. Bernard Minster, University of California, San Diego
Jill D. Pasteris, Washington University
Edward C. Roy, Jr., Trinity University

NRC Staff
Jonathan G. Price, Director
Thomas M. Usselman, Associate Director
William E. Benson, Senior Program Officer
Kevin D. Crowley, Program Officer
Bruce B. Hanshaw, Program Officer
Anne M. Linn, Program Officer
Lally A. Anderson, Staff Assistant
Charlene E. Anderson, Administrative Assistant
Judith L. Estep, Administrative Assistant
Shelley A. Myers, Project Assistant
Nathan L. Harshman, Research Aide

COMMISSION ON GEOSCIENCES, ENVIRONMENT, AND RESOURCES

The National Academy of Sciences is a private, nonprofit, self-perpetuating society of distinguished scholars engaged in scientific and engineering research, dedicated to the furtherance of science and technology and to their use for the general welfare. On the authority of the charter granted to it by the Congress in 1863, the Academy has a mandate that requires it to advise the federal government on scientific and technical matters. Dr. Bruce M. Alberts is president of the National Academy of Sciences.

The National Academy of Engineering was established in 1964, under the charter of the National Academy of Sciences, as a parallel organization of outstanding engineers. It is autonomous in its administration and in the selection of its members, sharing with the National Academy of Sciences the responsibility for advising the federal government. The National Academy of Engineering also sponsors engineering programs aimed at meeting national needs, encourages education and research, and recognizes the superior achievements of engineers. Dr. Robert M. White is president of the National Academy of Engineering.

The Institute of Medicine was established in 1970 by the National Academy of Sciences to secure the services of eminent members of the appropriate professions in the examination of policy matters pertaining to the health of the public. The Institute acts under the responsibility given to the National Academy of Sciences by its congressional charter to be an advisor to the federal government and, upon its own initiative, to identify issues of medical care, research, and education. Dr. Kenneth I. Shine is president of the Institute of Medicine.

The National Research Council was organized by the National Academy of Sciences in 1916 to associate the broad community of science and technology with the Academy's purposes of furthering knowledge and advising the federal government. Functioning in accordance with general policies determined by the Academy, the Council has become the principal operating agency of both the National Academy of Sciences and the National Academy of Engineering in providing services to the government, the public, and the scientific and engineering communities. The Council is administered jointly by both Academies and the Institute of Medicine. Dr. Bruce M. Alberts and Dr. Robert M. White are chairman and vice-chairman, respectively, of the National Research Council.

PREFACE

Drilling involves a set of processes for breaking and removing rock to produce boreholes, tunnels, and excavations. Drilling is used extensively for resource extraction, for building civil infrastructure systems such as sewers and subways, for environmental remediation, and for scientific purposes. Efficient and effective drilling technologies are critical elements in a robust and healthy economy. Improvements in the fundamental technologies applicable to the drilling of rock will benefit the U.S. economy and strengthen the competitive position of the United States in the worldwide drilling, excavation, and comminution industries.

The Geothermal Division of the Department of Energy is one agency of the U.S. government that hopes to find better and less costly ways of penetrating rock in order to harness geothermal energy resources more efficiently. With this goal in mind, the Geothermal Division asked the National Research Council to establish a committee to examine opportunities for advances in drilling technologies that would have broad industrial, environmental, and scientific applications such as energy exploration and production, mining, tunneling, water well drilling, underground storage, and environmental remediation. The formal charge to the committee is given in Appendix A.

The Committee on Advanced Drilling Technologies began its work in February 1993, and met four times over the course of the study. In April 1993 the committee invited 42 experts on drilling to a workshop that elicited ideas on advanced drilling technologies. A list of the invited experts and other participants is given in Appendix B. Results from this workshop assisted the committee in its assessment of the areas in which improvements are possible.

This report of the committee provides an examination of the technical and scientific feasibility of substantial advances in drilling and related technologies. In this report, the committee (1) examines concepts for new mechanical and nonmechanical drilling applications, including advances in the knowledge of tool-rock interaction; (2) identifies potential opportunities for research; and (3) makes recommendations on the scope and direction needed to realize these opportunities for improved methods of drilling.

The focus of the report is the physical systems used to create holes and tunnels in the subsurface. The report does not address other aspects of drilling related issues such as sample recovery and waste minimization. Although these are important issues, especially in environmental applications, they are outside the charge to the committee.

This study received direct support from the Department of Energy and the Gas Research Institute. The committee and staff gratefully acknowledge the support of each of these agencies.

CONTENTS

1
EXECUTIVE SUMMARY

Drilling involves a set of processes for breaking and removing rock to produce boreholes, tunnels, and excavations. In general, the object of drilling is to reach a target in the subsurface. The target may be a small feature at considerable depth or increasingly—particularly in petroleum industry applications—at substantial horizontal distances from the drilling site. The paramount objectives of drilling are to reach the target safely in the shortest possible time and at the lowest possible cost, with additional sampling and evaluation constraints dictated by the particular application.

Technology to drill holes and to excavate tunnels and openings in rock is vital for the economic and environmental well-being of the United States. During this century, U.S. technology has dominated the worldwide drilling industry and much of the excavation and comminution industries. In the committee's view, this U.S. dominance is likely to erode without continued technological advances.

With this concern in mind, the Geothermal Division of the U.S. Department of Energy asked the National Research Council to establish a committee to examine opportunities for advances in drilling technologies that would have broad industrial, environmental, and scientific applications such as energy exploration and production, mining, tunneling, water well drilling, underground storage, and environmental remediation. This report is the result of the committee's deliberations.

Drilling is a key technology in several applications of strategic or societal importance, including

- energy and mineral production,
- environmental protection, and
- infrastructure development.

Each of these applications would benefit from improvements in the basic **system** that breaks rock and removes debris from drill holes and excavations.

The drilling system comprises two principal parts: first, the mechanical, electrical, and hydraulic components, and second, the interactions between these components and the rock. Significant technological advances in drilling technology are likely to be obtained only through improvements in both aspects of the system. The goal of these technological advances, simply stated, is to drill and excavate safely at the lowest possible cost and in the shortest possible time.

There are both short- and long-term benefits from research and development (R&D) programs aimed at developing advanced drilling systems. Improvements in individual system components could be incorporated into conventional drilling systems almost immediately, providing short-term (less than five-year) payoffs. Longer-term payoffs will come from advances in basic R&D to assemble these individual components into a **smart drilling system**—a system capable of sensing and adapting to conditions around and ahead of the drill bit.

Recommendations

1. **R&D in advanced drilling technology is needed to improve the drilling system.** R&D should result in reduced costs and drilling times to more effectively achieve various drilling goals.

Drilling involves a complex set of mutually interacting, consecutive component operations (mechanical, hydraulic, and electrical) that must function in unison. An integrated systems approach is needed to ensure that these component operations function near peak performance with a minimum of discontinuities (failures in components that lead to system breakdowns). A *long-term* R&D effort could provide significant improvements in drilling technology through advances in understanding basic physical and chemical processes related to rock breaking and rock removal, and particularly through the development of flexible smart drilling systems incorporating improvements in sensing and guidance of autonomously advancing drilling units. Significant improvements in drilling technology could also be realized over the *short term* (less than five years) through incremental R&D on many of the rate-limiting processes and critical components of current drilling systems. A well-coordinated, incremental R&D effort could accelerate attainment of the primary goal of development of the smart drilling system.

2. **The principal thrust of an R&D program should be on the development of the smart drilling system.** A smart drilling system is one that is capable of sensing and adapting to conditions around and ahead of the drill bit in real time (i.e., while drilling), with minimal operator intervention. Such a system must be capable of assessing the mechanical properties of rock through measurement of physical and chemical (e.g., mineralogical) properties concurrent with drilling. The development of a smart system will require concerted technological advances in several areas, including the following:

- *Development of precise connections between measurable properties and local drilling resistance:* Such connections must be established based on the existing understanding of the connection between rock constitution and comminution mechanisms. Many physical and microstructural properties such as porosity, elastic properties, and wave attenuation can readily be measured locally. These must be associated more precisely with factors that govern the drilling resistance of rock through directed mechanistic studies and modeling to develop automated response characteristics of a smart system.
- *Development of sensors for the smart drilling system:* The smart drilling system requires sensors that are capable of detecting and measuring the following:

a. Conditions at the drill bit: Sensors are needed for in situ measurement of pressure (including pore pressure), temperature, permeability, mineralogic and chemical composition of the rock and heterogeneities, borehole fluid composition (at the part-per-million level for environmental applications), stress state, and rock strength.

b. Conditions ahead of the drill bit: Sensors are needed that measure rock properties (such as porosity, elastic properties, and wave attenuation) ahead of the drill bit to adjust drilling parameters, such as the weight on the drill bit and the rotary speed, and to avoid potential problems (e.g., blowouts or loss of circulation) while drilling.

c. Spatial position of the drill bit: Sensors are needed that are capable of detecting the position of the drill bit in space in order to steer the bit around undesirable zones and reach desired targets.

- *Development of control systems* for accurate positioning and steering of the drill bit and for automatically adjusting drilling parameters (e.g., load and torque on the drill string, flow rates of drilling muds and fluids), according to local conditions, is necessary to optimize rock breakage and rock removal. This will require precise information at the bit-rock interface of the rock breaking mechanism, rock strength, pressure, temperature, and stress state.
- *Development of improved methods for steering the drill bit:* A number of mechanical methods for steering are currently available, but large turning radii often preclude their application in a smart drilling system. To enhance steering capabilities, R&D is needed to develop downhole motors, flexible drill strings, and guidance techniques for smart drilling systems.
- *Continuous monitoring of the state of the entire drilling unit*, including wear of tools, state of other mechanical components, flow of coolant, and the like, is required to anticipate the occurrence of possible discontinuities.
- *Development of improved telemetry methods for transmitting real-time borehole data to the surface:* The use of advanced sensors for real-time, downhole measurements will require significant improvements in data telemetry. Such telemetry is essential for monitoring the smart system from the surface. At present, the most advanced telemetry systems utilize mud-pulse technology and are capable of transmitting data at only a few bits per second. Rates on the order of kilobits per second or higher will be required for advanced smart drilling systems. Telemetry is a rate-limiting step in present drilling systems, and it will become more so as smart drilling systems are developed.
- *Development of means for continuous and instantaneous support of the rock around the borehole:* The support provided by the rock itself should be used, where possible, in lieu of casing the hole as a separate operation.

3. Although the principal thrust of the proposed R&D program should be on the smart system, the program should also facilitate incremental improvements in all consequential aspects of present

drilling technology. This should result in more immediate attainment of greater efficiencies and cost savings. Such additional R&D should focus on the following problems:

- *Novel drilling technology, with a focus on the physics of rock removal to reduce energy requirements for drilling:* Hybrid systems that combine novel technologies to lower the drilling resistance of the rock and mechanical methods (e.g., fluid jets coupled with conventional rotary technologies) to break and remove the rock are especially promising. Initial attempts to develop combined novel-mechanical cutters should be conducted with mining or oil field bits because of their smaller size and lower cost. Once this technology is developed in drilling applications, the transfer to tunneling and excavating applications should be attempted.
- *Improved cutter materials and bearings:* Conventional drill bits have steel, diamond, or carbide cutters that remove rock by impact or shearing processes. New wear-resistant, diamond-coated cutters are finding increased use in hard, abrasive rocks. Advances in new wear-resistant materials are rapidly applied to cutters and high-speed bearings; as such, additional R&D on hard materials and their applications is encouraged.
- *Improved bits for drilling in heterogeneous materials:* Bits with polycrystalline or natural diamond cutters have the potential to drill much faster than conventional steel or carbide bits because they can operate at much higher rotary speeds and weight-on-bit loads. R&D is needed to utilize these wear-resistant cutter materials in multipurpose bits that can effectively drill through alternating layers of soft and hard rock.
- *Development of environmentally benign drilling fluids:* R&D is needed on the design of nontoxic drilling fluids and foams as alternatives to oil-base fluids, which may be both toxic and difficult to remove from the drillhole. These new drilling fluids must have (1) filtration control to minimize fluid invasion and damage to permeable zones and (2) lubrication to prevent differential-pressure sticking of the drill pipe against the borehole wall. They also must provide adequate hole-cleaning capabilities in horizontal and high-angle wells. Improvements in understanding the fundamentals of cutting transport, flow visualization, air/foam behavior, and fluid viscoelastic behavior will aid the development of such fluids.
- *Development of durable, compact, high-power downhole motors for directional and extended reach drilling:* Technology now exists to build downhole motors to increase drilling rates by factors of two to four by increasing the power delivered to the drill bit; the development of

such motors would be a short-term improvement. Additional areas for R&D include improved air-drilling motors and higher-power motors for hard-rock drilling.

4. This R&D program should be a national effort. Both the public and private sectors should benefit; resources and guidance for the program should be shared, where appropriate. This program should have the following characteristics:

- *Integration of industry, university, and government perspectives should be achieved.*
- *Federal support should serve primarily as a catalyst,* with industry providing both technological and financial support. The percentage of R&D support from the federal government and industry could be project specific. The actual R&D should be done by the best-qualified institutions whether in the private sector, universities, or government laboratories.
- *Finally, a long-term commitment* is needed to accomplish the objectives of the program.

The program should be structured with shared research objectives among the federal and industrial partners. Support of projects should be based on a peer-review process and assessment of how the results would contribute to overall program goals. Competition for research funds should be open to industry, national laboratories, and universities.

Attainment of the proposed enhanced drilling capabilities through both short-term and long-term R&D requires a long-range administrative structure that combines the discipline, mission orientation, and flexibility needed to nurture the required scientific and technological innovations.[1]

---

[1] Although the committee discussed a number of possible administrative structures, it ultimately concluded that recommendations in this area were outside its task and expertise.

2
ADVANCED DRILLING SYSTEMS

Introduction

Drilling involves a set of processes for breaking and removing rock to produce boreholes, tunnels, and excavations. In general, the object of drilling is to reach a target in the subsurface. The target may be a small feature at considerable depth, or increasingly—particularly in petroleum industry applications—at substantial horizontal distances from the drilling site. The paramount objectives of drilling are to reach the target safely in the shortest possible time and at the lowest possible cost, with additional sampling and evaluation constraints dictated by the particular application.

The Drilling System

The principal elements of current drilling systems are shown in Figure 2.1. The drill head comminutes rock at the end of the borehole. In most applications, the drill head is a drill bit that breaks rock by mechanical or mechanical-hydraulic action. The rate of breakage is governed by bit design, including the bit's effectiveness in breaking rock and resisting wear, and by rock type, temperature, pressure, and operating procedure (the experience and aptitude of the driller can have a considerable impact on how fast and effectively drilling proceeds). The drill is powered either from the surface through a drive string or by a downhole motor.

Rock breakage by the bit is followed by transfer of rock fragments to the surface. In tunneling applications, transport of waste material can involve elaborate mechanical systems; in petroleum and geothermal drilling, fragmented rock is typically transported to the surface by drilling fluid (mud) or air. The drilling fluid may provide power to drive the drill;

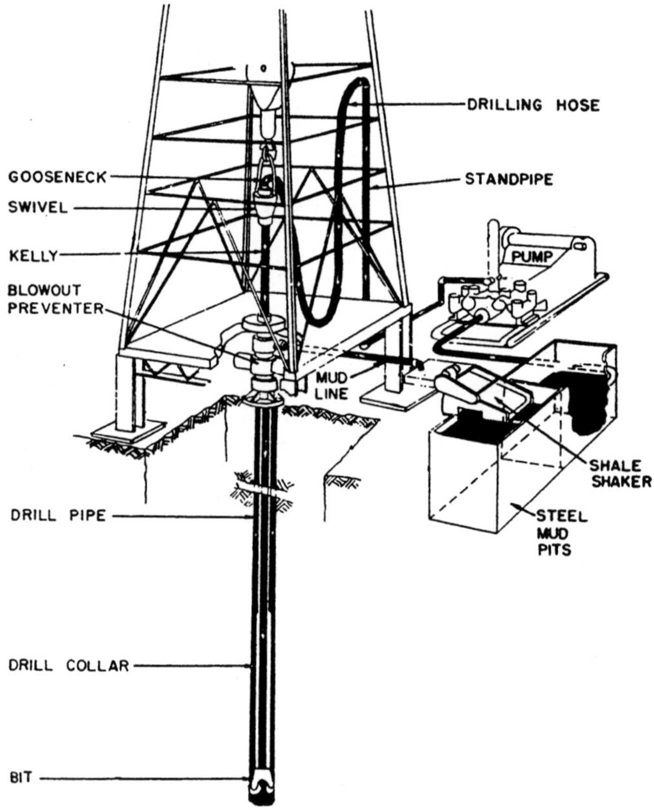

FIGURE 2.1 Principal components of current drilling systems (Bobo and Hoch, 1957).

it also acts as a coolant for the cutters, a conditioner for stabilizing the borehole (i.e., preventing borehole collapse and blowouts), and in some advanced applications, a medium for transmitting information to the surface by pressure pulses.

In most drilling systems, there is little or no downhole sensing of rock or bit conditions, and guidance systems, if present, are primitive. Although measurement-while-drilling and logging-while-drilling technologies exist, in most cases target acquisition requires interruption of the drilling process to insert special tools to obtain rock samples (core) or borehole measurements (e.g., pressure tests or well logs). Sensing the condition of the drill bit or wellbore commonly amounts to nothing more than an attempt (in many cases, unsuccessful) to recognize the warning signs of incipient loss of function or catastrophic failure.

Key processes of the typical drilling system can be subdivided into two groups (Table 2.1). The first group includes processes such as rock breakage, debris removal, and maintenance of borehole stability. These are the *rate-controlling processes* or *system bottlenecks*, and they limit the rates at which the other processes can operate. In present drilling practice, the rate and cost of a single process in this group may dominate the system. In the committee's view, improvements in these processes can be achieved through increases in their rates and reductions in their cost. Important, but evolutionary, improvements are possible through advances in these individual processes and in their integration.

The second group includes processes such as drill bit sensing (e.g., monitoring bit wear), rock properties sensing and evaluation, drill bit or drillstring steering, and wellbore damage sensing. Three factors make revolutionary advances in these processes possible and perhaps even likely: (1) these areas of drilling technology are relatively undeveloped, and therefore significant advances are possible; (2) these advances will be driven by the changing nature of drilling targets (e.g., smaller, deeper, harder to detect); and (3) this development will be facilitated by technological advances in related fields (e.g., computer science, microelectronics) that can be readily adapted to drilling. As described in the following chapters, important recent advances in sensing while drilling and directional drilling show that rapid and striking advances are already under way.

This report focuses on the development of a *smart drilling system*—a self-guided drilling system that has the potential to operate in a way that permits rapid, efficient, and damage-free detection and acquisition of targets without costly breakdowns and discontinuities.

Smart Drilling Systems and the Systems Approach

The *smart drilling system* is a system capable of sensing and adapting to conditions around and ahead of the drill bit to reach desired targets. This system may be guided from the surface, or it may be self-guided, utilizing a remote guidance system that modifies the trajectory of the drill when the parameters measured by the sensing system deviate from expectations.

TABLE 2.1 Key Elements of Drilling Systems and Areas of Possible Evolutionary and Revolutionary Improvement

DRILLING SYSTEM PROCESS	CURRENT STATUS	ANTICIPATED LEVEL OF IMPROVEMENT
Rock breaking	Key element in drilling process: bottleneck to increased drilling rate	Evolutionary
Debris removal	Potential bottleneck, especially in tunneling	Evolutionary
Borehole stabilization	Discontinuous process	Evolutionary
Drill bit sensing and evaluation	Technology not available	Revolutionary
Rock properties sensing and evaluation	Some measurement-while-drilling capability now exists	Revolutionary
Drill bit positioning and steering	Notable recent advances in steering	Revolutionary
Borehole sensing	Technology not available	Revolutionary

The smart drilling system does not currently exist, but it is presaged by recent dramatic advancements in directional drilling and in technologies of measurement while drilling. Rapid innovation in microelectronics and other fields of computer science and miniaturization technology holds the prospect for greater improvements—even revolutionary breakthroughs—in these systems.

The development of smart drilling systems has the potential to revolutionize drilling. Research in this area will have a significant impact on drilling success and overall cost reduction. Such "smart" systems are increasingly needed to overcome the drilling challenges posed by small, elusive, easily damaged subsurface targets. This is particularly true in applications where the identification of small or difficult-to-predict drilling targets and formation damage are key issues in drilling success.

In the development of smart drilling systems, the required improvements in the sensing elements of the system will have an impact on other processes such as rock breaking, debris removal, and borehole stabilization. Revolutionary advances in these fundamental processes might be possible as information about the subsurface environment becomes available in real time.

The most effective means of operational implementation of the smart drilling systems and all other secondary goals recommended by this committee is the adoption of an *integrated systems approach* to all major drilling practices. In the integrated systems approach, all component parts and processes are designed to function in unison at an optimum level of performance without excessive redundancies and without overloading any of the interdependent component parts and processes. At present, this is rarely the case in drilling. The driller must adjust drilling parameters to changing conditions based on very limited information—with some modifications guided primarily by experience and intuition. Further, a host of discontinuities can and do interrupt these processes. These discontinuities include (1) tool-bit wear; (2) degradation and loss of effectiveness of the drilling fluid; (3) damage to the wellbore (e.g., borehole breakouts); and (4) time out for well testing and wireline geophysical logging. Discontinuities are costly and time consuming and, in the worst case, catastrophic (e.g., lost circulation, borehole collapse). Adherence to a systems approach in all instances should reduce costly discontinuities to acceptable levels.

References

Bobo, R. A., and Hoch, R. S., 1957, Keys to successful competitive drilling - part 5-A: World Oil, v. 145 (4), p. 91-98.

3
NATIONAL IMPORTANCE OF DRILLING

Introduction

Drilling is a key technology in several applications of strategic or societal importance, including

- exploration for and extraction of oil, gas, geothermal, and mineral resources;
- environmental monitoring and remediation;
- underground excavation and infrastructure development; and
- scientific studies of the Earth's subsurface.

Drilling is the primary tool for extracting petroleum from rocks in the subsurface. Improvements in drilling technology that lower drilling costs and increase the rate of success in finding and extracting petroleum will have a direct benefit to the United States in terms of higher energy reserves, stable energy costs, and improved economic competitiveness in the drilling and service industries, which are increasingly global in character. Drilling is also the primary tool for extracting geothermal energy (hot water and steam) from the subsurface for heat and electricity production. At present, geothermal energy is more expensive than fossil fuel energy, owing in part to the high cost of drilling. The reduction in drilling costs through the introduction of improved technologies will allow more of this clean, domestic energy source to be utilized.

Drilling is becoming an increasingly important tool for environmental protection and remediation. Drilling is a relatively noninvasive method for investigating and removing chemical and radioactive wastes from the subsurface, and for placing barriers in the subsurface to halt the spread of contamination. Improvements in drilling technology will improve the efficiency of waste extraction and thereby lower the cost of cleanup efforts.

Drilling technology, including tunneling technology, is finding increased application in the development of urban infrastructure (utilities, transportation, and communications facilities), much of which is located underground. A significant fraction of this development is supported directly or indirectly by taxpayers. Improvements in drilling technology will lower costs and could allow more infrastructure to be located underground, thereby increasing aboveground living space in urban areas.

Oil Drilling

The U.S. petroleum industry maintains a high level of drilling and spending for drilling-related goods and services in the continental United States, despite the oil price collapse in 1986. This substantial level of activity should continue in the foreseeable future. In 1990, total petroleum industry exploration and production was $45.2 billion (American Petroleum Institute [API], 1991). Expenditures for drilling comprised about $10.9 billion of this total and were concentrated mostly in exploration and development well drilling.

Although the U.S. petroleum industry continues to maintain a considerable level of domestic drilling activity, the character of the companies drilling domestic petroleum wells is changing. Major petroleum companies are redirecting their exploration and production budgets for work abroad, and drilling activity by these companies in the lower 48 states has reached historically low levels. An increasingly large number of domestic wells are now drilled by small- to moderate-sized companies (independents) rather than major companies. Independents drill about 85% of domestic exploratory petroleum wells in the United States (Bode, 1992). Many independent companies occupy niches in certain geographic or technological areas in order to reduce costs and increase success rates. In the future, domestic oil and gas development will probably be dominated by these technically oriented, small- to moderate-sized companies or by major companies that have decentralized activities (Fisher, 1993). This change in the character of the industry reflects in part the changed character of the remaining resource base.

Statistics compiled by the API (Table 3.1) indicate the magnitude of the domestic petroleum oil and gas industry drilling effort. In 1992, the U.S. petroleum industry drilled 124,148,449 ft in 23,998 wells (World Oil, 1993). Oil well drilling continues to slightly outpace gas well drilling.

TABLE 3.1 Estimated Wells and Footages Drilled by the Petroleum Industry in 1992 (World Oil, 1993)

CONTINENT	OIL	GAS	DRY	OTHER	TOTAL	FOOTAGE
North America	10,981	8,887	7,875	902	28,745	144,436,169
United States	8,596	7,929	6,610	863	23,998	124,148,449
South America	1,462	62	274	192	1,990	12,477,671
West Europe	256	189	170	216	831	7,534,692
East Europe	9,965	786	860	88	11,699	84,797,571
Africa	408	36	99	66	609	5,670,955
Middle East	256	23	63	51	1,033	7,096,678
Far East	7,323	235	250	2,265 [a]	11,149	61,530,591
South Pacific	61	41	94	11	207	1,465,770
World	30,712	10,259	9,685	3,891	56,263	325,010,097

[a]Mostly mainland China.

In 1992, the number of gas wells drilled was 7,929, compared to 8,596 oil wells. Drilling between depths of 1,250 and 9,999 ft accounted for most oil well drilling in 1992; the average depth for all wells in 1992 was 5,444 ft.

The number of wells drilled by the petroleum industry in the United States amounted to about 43% of wells drilled worldwide in 1992, and U.S. drilling accounted for about 68% of all dry holes drilled worldwide in 1992 (Table 3.1). This reflects the increasingly complex and elusive nature of potential reservoirs, as well as limitations of current methods for locating hydrocarbon deposits. The proportion of wells drilled and the total drilled footage outside North America should increase as additional areas around the world are explored and developed. Complex geological conditions and difficult geographic circumstances in many of these areas will require the capability to remotely sense conditions in the subsurface and to drill holes in different orientations.

In 1990, the average cost for "conventional wells" (i.e., vertical wells drilled by using standard equipment) was about $75/ft (API, 1991). Drilling costs reflect the depth, type, and location of wells and the costs of drilling-related services. A comparison of wells drilled to similar depths at similar locations indicates an actual decrease in drilling costs for conventional wells since 1984 (API, 1991). This likely reflects, in part, improved drilling efficiencies and lowered costs resulting from advances in technology.

Petroleum well drilling in the United States is essential to ensure a stable domestic supply of energy. Recent estimates of the volume of recoverable resources suggest that operators should be able to improve their ability to add more reserves per unit of effort through improved technology (Fisher, 1993). In 1992, a panel of oil resource analysts convened at the request of the U.S. Department of Energy concluded that there is a substantial remaining, recoverable volume of crude oil in the United States, on the order of 99 to 204 billion barrels (Oil Resources Panel, 1992). The range in estimates reflects different assumptions of price and technology. This recoverable resource is the target of drilling for oil resource development (Fisher, 1987; U.S. Department of the Interior, 1989).

Although the undiscovered oil resource base is substantial in the aggregate, it is different from what has been pursued historically. The nature of the typical oil drilling target is changing from large reservoirs to smaller, less readily detectable reservoirs (Fisher, 1988). The fields being developed today generally are more geologically complex; consequently, their development requires reserve growth drilling.[1]

In summary, the potential oil resource accessible by drilling is significant, and U.S. oil and gas drilling for this resource will continue at or near current levels for at least the foreseeable future. The potential oil resource accessible by drilling is significant. However, reservoirs are increasingly smaller in size and are located in more geologically complex settings. Advances in drilling technology are especially important for accessing these targets and reducing overall development costs.

Natural Gas Drilling

Natural gas, which is composed mostly of methane, is a relatively clean and domestically abundant fuel that provides more than one-fifth of the primary energy used in the United States. Natural gas is particularly important in the residential sector, where it supplies nearly half of the energy consumed in U.S. homes (Energy Information Administration, 1993). It can also be liquefied for use in transportation as compressed natural gas.

There is potential for growth in the use of this fuel for transportation and for generation of electricity, given the large size of gas reserves in the United States and the availability of sophisticated natural gas production and delivery systems. The use of natural gas to generate electricity is projected to increase from 2.8 trillion cubic feet (Tcf) in 1992 to 5.5 Tcf by the year 2005 (U.S. Department of Energy, 1992), primarily due to the relatively clean-burning nature of natural gas compared to other fossil fuels.

At present, most natural gas is produced in the lower 48 states from conventional sources (Table 3.2). In 1991, about 20.5 Tcf of gas were

---

[1] Such as infill drilling, intrapool recompletions, or horizontal drilling.

TABLE 3.2 Current and Projected U.S. Gas Supply (Tcf or quads, see footnote 2)—Prices (1992$ per 10^6 Btu) are shown in square brackets (Woods, 1993)

AREA	1991	2000	2010
Lower-48 states	18.0 [$1.52]	19.4 [$2.26]	21.0 [$3.14]
Alaska[a]	0.4	0.4	0.5
Imports[b]	1.8 [$1.78]	2.8 [$2.43]	3.8 [$3.27]
Nonconventional	0.2 [$3.44]	0.1 [$3.54]	0.1 [$4.15]
Totals	20.5 [$1.56]	22.7 [$2.29]	25.4 [$3.16]

[a] Prices were not developed for gas production consumed in Alaska or exported.
[b] Imports have a higher price because they often enter the transmission system downstream from the wellhead.

consumed domestically at an average price of about $1.56 per 10^6 Btu.[2] Gas consumption is expected to increase to about 25.4 Tcf per year by 2010 at an average price of $3.16 (in 1992 dollars) per 10^6 Btu. The majority of this supply (21.5 Tcf) is expected to be produced in the lower 48 states and Alaska.

Recent domestic supply projections indicate that a substantial resource of about 1,300 Tcf exists in the lower 48 states (National Petroleum Council, 1992; Enron Corporation, 1993). At present rates of U.S. gas consumption (approximately 20 Tcf/yr), this amounts to a supply of about 65 years. This resource includes 160 Tcf of proved reserves, 616 Tcf of conventional resources, and 519 Tcf of unconventional resources (National Petroleum Council, 1992). Major unconventional sources include

---

[2] 1 Tcf of dry gas has the energy equivalent of 1×10^{15} British thermal units (Btu), or 1 quad (1 quad = 1×10^{15} Btu).

low-permeability (tight) gas in shale and sandstone, and gas found in coal seams (coalbed methane). For these sources, technological advances may be particularly critical to efficient exploration and development because reservoirs are difficult to detect and access. This is also true for conventional gas resources located at depths greater than 15,000 ft (deep gas). Of the more than 2 million wells drilled to date in the United States, only about 16,000 reached depths of 15,000 ft or greater. Tight gas and deep gas are likely targets for targeted drilling, directional and horizontal drilling, and smart drilling technology.

Geothermal Drilling

Geothermal energy, including steam, hydrothermal, hot dry rock, and magma resources, constitutes a large and relatively untapped source of energy in the United States. The U.S. Geological Survey estimates total geothermal resources in the upper 10 km of the Earth's crust in the United States to be between 210,000 and 1,100,000 quads (Table 3.3), which is several orders of magnitude larger than current annual rates of domestic energy consumption.[3] Rex and Howell (1973) estimate that geothermal energy in the United States could supply 400,000 megawatts (MW) of power for a projected life of 100 years.[4] The U.S. Department of Energy (1991) estimates that geothermal energy output will rise from 20 billion kilowatthours (kWh) in 1990 to 184 billion kWh in 2030 and will account for about 3% of all U.S. electricity generation.

Geothermal energy is used in the vicinity of the production site for heating purposes (e.g., space heating and chemical processing) or is converted to electricity for long-distance transport, in which case it competes directly with oil, gas, coal, and nuclear energy. The cost of geothermally produced electricity exceeds that for most other energy

---

[3] The current annual rate of domestic energy consumption is approximately 82 quads (Energy Information Administration, 1993).

[4] In 1993, the United States had 2,725 MW of installed capacity of geothermal energy from plants in California, Hawaii, Nevada, and Utah (U.S. Department of Energy, 1993). For comparison purposes, U.S. electrical generating capacity (i.e., net summer capability) in 1992 was approximately 695,000 MW (Energy Information Administration, 1993).

sources. For example, current electricity prices are \$0.025 to \$0.060 per kWh in California (Kito, 1993), compared to Standard Offer 4 prices for geothermally produced electricity of \$0.10 to \$0.11/kWh.[5] In allowing for geothermal power plant costs, which are about \$0.03 to \$0.04/kWh, steam must be delivered to the power plant for less than \$0.03/kWh to be directly competitive at current prices, compared to \$0.07/kWh under Standard Offer 4 prices. At current electricity prices in California, it is not economical to build a geothermal power plant or to purchase geothermal energy.

TABLE 3.3 Estimated Geothermal Resources in the Upper 10 km of the Earth's Crust in the United States (Muffler, 1979)

SOURCE	ESTIMATED RESOURCE (quads)
Hydrothermal	110,000
Hot dry rock	50,000-500,000
Magma	50,000-500,000
Total	210,000-1,100,000

Geothermal energy is more costly than oil and gas at current prices, owing to the relatively low energy capacity of geothermal water and steam. In general, high numbers of large-diameter wells must be drilled in order to obtain sufficient water and steam for heating or electrical production. Geothermal reservoirs are frequently found in mixed hard and soft volcanic rocks, which makes drilling difficult and costly. Only a small portion of known geothermal resources can be exploited at current energy prices. Exploitable resources are usually found at shallow depths (generally less than 4,000 ft) and have extremely high productive capacities.

---

[5] Standard Offer prices are established by the California Public Utilities Commission for the purchase of electricity from qualifying facilities.

There is considerable room for improvement in technology for geothermal drilling. Improvements are needed to increase penetration rates and tool life in the mixed (hard-soft) volcanic rocks that form many geothermal reservoirs. Improvements are also needed to reduce the number of "discontinuities," such as lost circulation and stuck pipe, that typically occur while drilling in geothermal reservoirs. These improvements will allow deeper or less productive geothermal resources to be exploited.

Environmental Drilling

The cost of cleaning up hazardous wastes in the United States may exceed $1.2 trillion to be spent over the next 30 years, with a "best guess" value of $750 billion. Of this, about one-third ($240 billion) will be spent on the government's weapons complexes (Nuclear Waste News [NWN], 1991). The market for environmental consulting may grow at a rate of 20% per year from its 1991 value of $8.2 billion (NWN, 1992a), and simultaneously, the market for environmental remediation work will grow from about $2.5 billion in 1992 to more than $5 billion in 1995 (NWN, 1992b).

There are approximately 45,000 sites across the United States that are in some way contaminated by radioactivity; half of these are owned by the government (NWN, 1992c). Perhaps the best-known of these is Hanford (Washington), where there are 149 single-shell and 28 double-shell storage tanks for radioactive waste ranging in size from 55,000 to 1 million gallons (NWN, 1992d). Other sites at Hanford contain smaller amounts of radioactive material, as well as nonradioactive pollutants such as chlorinated hydrocarbons.

The number of U.S. sites at which hazardous but nonradioactive pollution occurs is not known with certainty, but it is probably at least as large as the number of radioactive sites. In a majority of cases, pollution is in the form of solid or liquid wastes that have escaped into the subsurface. Depths of penetration range from the immediate surface to at most a couple of hundred feet. Cleanup will involve two stages, namely, site characterization followed by remediation. No techniques are yet available that allow the extent of subsurface pollution to be estimated noninvasively. Further, remediation will always require access to the polluted zone. Drilling is probably the least disruptive method presently available for examining the subsurface. Remediation methods that operate through one or more boreholes will certainly be preferable to, for example, excavating the entire site and removing the polluted ground for treatment elsewhere.

The fraction of total remediation costs attributable to drilling is not easily estimated. The difficulty of access, the type of pollutant, and the nature and time of treatment will influence the total cost. In petroleum operations, drilling costs typically account for 50 to 80% of exploration finding costs, and about 30 to 80% of subsequent field development costs (Vincken, 1987). Typical costs for shallow hydrocarbon wells (up to 1,250-ft depth) drilled in the United States are about $27/ft (Anderson and others, 1991).

Whether this figure applies to typical environmental projects is not clear, but the sequence of operations—namely, identification of the zone of interest, test drilling, "production" drilling, and pumping operations to "produce" fluids—is of similar relative importance. The boreholes required for environmental remediation will be shallow, so it might be expected that they will cost in the range of $20 to $30/ft, similar to shallow petroleum wells. However, special circumstances may increase these costs substantially. If the drilled solids contain toxic or radioactive substances, the cost of drilling may increase dramatically because of the need to collect, document, and dispose of the cuttings and to decontaminate drilling equipment. At the Hanford site, for example, typical drilling rates (cable tool) are about 8 ft/day, at a cost of $800 to $1,000/ft (Volk, 1992). Alternative drilling methods are under investigation at the Hanford site, including the use of resonant sonic drilling methods (Volk, 1992; Volk and others, 1993) and the cone penetrometer (NWN, 1992e). The former has about twice the rate of penetration and is similar in cost to the cable tool method (Volk, 1992). Cost estimates for the cone penetrometer are as low as $50/ft (NWN, 1992e).

The cleanup of small-scale, nonradioactive subsurface contamination (e.g., local gasoline spills) will probably be more straightforward, and it is expected that drilling costs will be much lower. In the absence of any estimate of the relative numbers of wells that will be drilled in "difficult" and "easy" circumstances, a very approximate estimate can be obtained by assuming that the fraction of the environmental budget spent on drilling will be the same as the fraction of a petroleum operations budget spent on drilling. If this estimate is correct, then by using the figures given previously, the total value of the environmental drilling market ranges from $225 billion to $960 billion, with $72 billion to $192 billion to be spent on the government's weapons facilities. In any case, it is clear that there is a substantial opportunity for drilling research to develop techniques for drilling in contaminated ground, with a major issue being the handling of contaminated cuttings.

Once a contaminated site is characterized, remediation wells have to be drilled to place barriers, to inject or pump out subsurface fluids ("pump and treat"), or to use air sparging and other remediation activities. Many of these wells will have to be directionally drilled. At the Department of Energy Savannah River Site, several directional and horizontal wells have been drilled (Kaback and others, 1989a, b; Westinghouse Savannah River Co., 1992). Early work was carried out by using conventional oil field directional drilling technologies, but subsequent holes have been drilled with purpose-built equipment or with drilling techniques adapted from river-crossing and service placement technologies. The use of horizontal wells to remove contaminants from the Savannah River Site saved $125 million compared to a conventional pump-and-treat program (NWN, 1992f).

Once the hole has been drilled, it is not always easy to keep it open, particularly in very soft, sandy ground. Various consolidation techniques are available from the oil industry and from civil engineering practice, but many rely on the use of grouts[6] that themselves introduce undesirable additional pollution. One promising technique that is under investigation (Simon and Cooper, 1994) is to freeze the ground around the borehole, thus preventing collapse until a protective casing can be set.

An application of particular importance in the field of environmental characterization is the need to sample subsurface solids and fluids accurately. In the case of solids, issues concern the requirement to recover undisturbed cores so as to identify chemical species that are present and also to measure strength, porosity, and permeability. In the case of the fluids, the nature of the fluids, dissolved solids concentration, and saturation are important parameters.

In taking a longer perspective, there is an obvious need for directional drilling instrumentation and steering techniques that will be able to monitor and react to the surroundings as the hole is drilled. This would allow the detection and avoidance of undesirable obstacles such as pipelines, tanks, or contaminated regions. It would also allow the drilling assembly to be

---

[6] Cement slurries that are injected into the subsurface, usually through boreholes, to seal fractures and other openings.

steered to follow a desired direction, for example, to stay at or near the water table or to follow permeable layers.

In summary, there is a large market for "environmental drilling." It is characterized by a need for minimally invasive vertical (investigation and monitoring) to horizontal (remediation) drilling at relatively shallow depths (maximum, 300 ft). Penetration of loose ground is probably the most difficult problem, because of poor hole stability, particularly if the ground contains a mix of hard and soft material (e.g., sand and boulders). Although existing drilling technologies may work under some conditions, there is no universally applicable technique. For some applications, there is no acceptable method available at all.

Service Companies

U.S. service companies have historically been leaders in drilling technology related to the oil, gas, mining, and tunneling industries. This leadership is being threatened by economic problems and the fact that mining and oil companies are moving overseas. Data from the Bureau of Labor Statistics (Fletcher, 1992) show that from the peak drilling activity in 1981, the oil and gas industry has lost more than 369,000 jobs. Total annual oil and gas exploration and production expenditures paid to service companies decreased from $36.7 billion in 1981 to $10.6 billion in 1990 (Independent Petroleum Association of America, 1991). Job losses in the oil and gas service companies include many of the technical leaders in this field.

Service company technical leadership was established and has been maintained for several reasons. Until recently, there has been a large domestic market for these companies, because of the large number of holes drilled for oil and gas in the United States. The profitability of the service industry was high, which allowed it to make large expenditures for R&D. This industry also had the benefit of an excellent U.S. educational system, which turned out large numbers of trained engineers, as well as excellent technical resources, including universities and national laboratories.

The world leadership of U.S. service companies has been diminished by several recent developments. First, and foremost, is the reduction in the exploration and production of domestic mineral resources, due mainly to a reduction in market prices for petroleum and certain mineral commodities, as well as environmental restrictions. There has been a concomitant reduction in profitability of the major petroleum, mining, and service companies. These companies have slashed their budgets and pared their staffs, and many have significantly reduced or eliminated their spending for R&D. Overseas operations of domestic service companies have also been affected by foreign laws and policies that exclude them from certain markets.

The leadership of U.S. service companies is also being challenged by foreign government-subsidized projects to develop advanced drilling tools. Examples of such projects include (1) extended reach offshore drilling tools (6 to 8 miles) to drain entire offshore fields from single platforms (BP and Statoil—North Sea); (2) slim-hole drilling systems to significantly reduce drilling costs in remote areas (BP and Shell); (3) deep-water offshore oil production systems (Petrobas—Brazil); (4) hard-rock geothermal drilling systems (Komatsu—Japan); and (5) advanced deep-well guidance systems (KTB—Germany). U.S. service companies do not presently have the resources to undertake these types of long-term development projects.

In the committee's view, development of advanced drilling systems is needed to keep U.S. service companies competitive. These advanced systems will require the integration of advanced mechanical, computer, hydraulic, electronic, and rock destruction technologies. U.S. service companies are specialized and therefore do not have the in-house capabilities to develop all components of these systems.

Infrastructure, Underground Excavation, and Mining

Infrastructure and underground excavation include facilities such as water and sewers, railroads, highways, mass transit, and communications. Demands for new facilities are increasing at an accelerated rate due to the increase in urban populations. In addition, there is a high demand for replacement facilities because many existing facilities have exceeded their design lives. Until recently, most of these facilities have been located on or near the surface. However, with increased urban crowding, there is increased interest in building these replacement facilities underground.

Recent estimates suggest that over 200 miles of tunnels, 5 ft and larger in diameter, will be built in the United States by the end of the century (American Underground Space Association, 1993). In addition, approximately 1,000 to 3,000 miles of microtunnels, tunnels between 1 and 5 ft in diameter, and perhaps another 10,000 miles of tunnels less than 1 ft in diameter, are expected to be built in the same time period (Kramer and others, 1992; Boyce, 1993; Iseley, 1993). This represents a capital expenditure of $7 billion to $10 billion for driving the tunnels alone. In contrast to drillholes discussed earlier, these tunnels are one to two orders of magnitude greater in diameter, they are generally horizontal, and they must be lined to provide a service life of 50 to 100 years. Additionally, these tunnels must be driven through a variety of geologic materials from soft, water-bearing silts and clays to very hard bedrock, and very often they must be driven in crowded and environmentally sensitive urban areas.

The owners and users of infrastructure and underground facilities generally are the taxpayers, through federal, state, and local governments. In some instances (e.g., communications and some utilities) the immediate owners are regulated or quasi-governmental entities, but the ultimate users are taxpayers. In the past, the federal government often provided the majority of funding (up to 90%) for construction of these facilities. However, the current and continuing trend is toward greater financial participation of local governments through taxation, bonds, or user fees.

In the privately owned mining industry, much of the exploratory drilling is similar to that described previously for the oil and gas industries. In addition, the mining industry uses boring machines to gain access to ore bodies and to drive operations connections between various areas of mines. These machines are similar to, or use technology similar to, tunnel boring machines (TBMs) used for infrastructure and underground excavation. The mining industry also uses smaller machines to drive smaller connecting tunnels and shafts, and to provide communications and utility connections between various areas of mines. These machines are similar to, or use technology similar to, microtunneling machines. Estimates for the lengths of TBM- and microtunneling-like excavations in mining operations are not available. Discussions with Professor L. Ozdimir, Colorado School of Mines, have led to the rough estimate that by the end of the century, these might total 100 miles of mining tunnels greater than 5 ft in diameter and 500 miles of microtunnels (Ozdimir, 1993).

In the past, improvements in tunneling technology have followed an evolutionary path, owing to gradual changes in machine design by TBM manufacturers and general contractors. When TBMs were introduced some 35 years ago, tunneling rates were measured in feet per day at best; tunneling rates of modern machines under ideal conditions are measured in hundreds of feet per day. Advances in U.S. TBM technology have slowed, mainly because domestic manufacturers are leaving the business or are being acquired by foreign firms. In the committee's view, the ability of U.S. industry to build next-generation TBMs will be difficult without a coordinated domestic effort. A coordinated effort is needed for TBMs because although some components can be tested individually, the machine itself must be tested by prototype. Only stable, well-capitalized companies can handle the risk associated with the development of new technology, given the expense of prototype design, construction, and testing.

Continued improvement is expected in tunnel construction technology, but it is also expected that, left to the continuing historical evolutionary process, advances will continue at an average rate of only a few percent per year. A coordinated development program would greatly accelerate this rate of improvement by concentrating resources and expertise.

Drilling for Scientific Purposes

Drilling provides a vital operational avenue to satisfy a multitude of scientific purposes, both on the continents and in the oceans. Although drilling and exploration for resource recovery remain important purposes, the overall objectives of scientific drilling are far broader and include the following (National Research Council [NRC], 1979, 1988, 1992):

- structure and chemical constitution of continental crust;
- distribution of mineral resources;
- thermal regime of the crust and crustal heat flow;
- state of stress in the Earth's crust and crustal response to stress, including properties of fault zones for purposes of understanding earthquake phenomena; and
- nature and age of the ocean floor with particular reference to seafloor spreading.

A considerable amount of scientific drilling is supported by federal agencies in order to address several societally important issues, such as supplies of energy, water resources, mineral resources, environmental management, disposal or storage of hazardous wastes, siting of dams, and location of nuclear power stations. In 1978, federal expenditures for continental drilling amounted to $500 million/yr (NRC, 1979). Although the portion of this total designated purely for scientific purposes was much smaller, scientific investigations of various magnitudes were associated with nearly all such drilling activity. In more recent times, for example, from 1985 to 1992, the extent of scientific drilling in the continental United States involved a total of 31 holes to an average depth of 425 m, with one reaching a depth of 3,250 m and a temperature of 350°C. The purposes of these programs included activities such as the study of crustal structure (5 holes), volcanology (4), geothermal processes (21), and fault tectonics (1). The present (1992-1993) scientific drilling activity in the United States involves about 20 projects reaching an average depth of 902 m for rather similar purposes such as the study of sedimentary basins (7 holes), geothermal processes (4), volcanology (4), mineral deposits (2), and meteorite impacts (3). In addition to these, the United States is an important collaborator in several large international scientific drilling programs such as the Lake Baikal, Kola Peninsula, and KTB deep drilling programs (MacGregor, 1993).

In distinction to the continental drilling programs, the Ocean Drilling Program of recent years involves an outlay of roughly $40 million/yr but is devoted to almost exclusively to scientific purposes (NRC, 1992).

There are well-established organizational means of exploiting most drilling opportunities for scientific purposes. Several recent NRC studies of both continental drilling (NRC, 1979, 1988) and ocean drilling (NRC, 1992) have assessed the goals, potentials, and successes of these programs and have recommended well-thought-out, long-range operational procedures for nurturing, selecting, and funding them. These studies have concluded that in a number of important areas, key improvements in the state of the art of drilling could markedly improve the attainment of scientific objectives (NRC, 1979, 1988). These objectives include drilling to greater depths (~10 km), drilling through harder rock at higher temperature (~500°C), and sample collecting from these environments, which often have higher reactivity. Other recommendations include development of long-term fluid-flow measurement techniques; techniques for slim-hole drilling with continuous casing capability; better techniques to determine the in situ state of stress in the crust; and development of means for making reliable long-term incremental stress measurements in such adverse environments.

In areas more specifically related to hydrocarbon resource recovery, recommendations have included improved a priori reservoir evaluation to increase the fraction of available resources that can potentially be recovered (the average present recoverable fraction in most oil fields is no better than one-third, with the remainder being declared nonrecoverable [NRC, 1988]). This requires the development of better predrilling geophysical assessment of potential reservoirs. Even with the above provisos, however, the long-range objectives associated with scientific drilling are generally being met with present technology.

In many areas, the demands of scientific drilling programs in terms of (1) precision and depth in hard rock, as in the KTB Scientific Bore Hole program (Zoback and others, 1993a); (2) required sophistication for core recovery, as in the projected San Andreas Fault Drilling Project (Zoback, 1993b); and (3) overall complexity in undersea drilling at high pressures and temperatures into hard rock (NRC, 1992) both act as stimulus for radical improvements in drilling technology and provide the opportunity for such developments.

Because the cost of drilling is usually the major part of any scientific drilling program, such programs require careful planning to utilize all available opportunities, both nationally and internationally. Therefore, the scientific drilling community must be well organized and fully cognizant of developments in commercial drilling. Alternatively, in any long-range development plan for radical improvements in drilling for resource recovery, full liaison with the scientific drilling community is essential.

Summary and Recommendations

Drilling is a key technology in several applications of strategic or societal significance, including:

- exploration for and extraction of oil, gas, geothermal, and mineral resources;
- environmental monitoring and remediation;
- underground excavation and infrastructure development; and
- scientific studies of the Earth's subsurface.

All of these applications would benefit from improvements in the basic drilling system that breaks rocks and removes debris from boreholes and excavations. Such advances should include the following:

- increasing rates of penetration and tool life through improvements in cutter technology and materials;
- improving capabilities to sense conditions at and ahead of the tool in order to locate targets or avoid obstacles in the subsurface; and
- improving the ability to steer the bit and to drill directional, or horizontal holes to reach desired targets or target zones.

The objective of these improvements is to enhance the ability to locate and reach targets in the subsurface and to reduce the overall costs of doing so. Indeed, these improvements in drilling technology will enhance the competitive position of U.S. companies in several sectors of the economy, most notably in the areas of resource exploration and extraction, environmental remediation, and infrastructure development.

The committee believes that improvements in drilling technology can best be achieved through a national R&D effort that integrates industry, university, and government perspectives. Federal support for this effort should be used primarily as a catalyst, with industry providing both technological and financial support. The actual R&D should be done by the best-qualified institutions, whether in the private sector, universities, or government laboratories, with the percentage of R&D support from the federal government and industry being project specific.

The committee also believes that the program should be structured with shared research objectives between federal and industrial partners. Support of projects should be based on a peer-review process and assessment of how the results would contribute to overall program goals. Competition for research funds should be open to industry, national laboratories, and universities.

Attainment of the proposed enhanced drilling capabilities through both short- and long-term R&D requires a long-range administrative structure that combines the needed discipline, mission orientation, and flexibility to nurture the required scientific and technological innovations. The committee discussed a number of possible administrative structures, but it ultimately concluded that recommendations in this area were outside its task and expertise.

References

American Petroleum Institute, 1991, Joint Association Survey on 1990 Drilling Costs: Independent Petroleum Association of America, Mid-Continent Oil and Gas Association, 106 pp.

American Petroleum Institute, 1993, Quarterly completion report, v. 9 (1), 65 pp.

American Underground Space Association, 1993, News, v. 8 (3), p. 27-30.

Anderson, E. E., Cooper, G. A., Maurer, W. C., and Westcott, P. A., 1991, An analysis of relative costs in drilling deep wells: SPE 22574, Proceedings of the Society of Petroleum Engineers 66th Annual Technical Conference and Exhibition, Dallas, Tex., Richardson, Tex., SPE, p. 355-364.

Bode, D., 1992, The best of times and the worst of times and back: Presentation to the Independent Petroleum Association of America (IPAA), London, England, November 17, 1992, 12 pp.

Boyce, G., 1993, Personal communication.

Energy Information Administration, 1993, Annual Energy Review 1992: U.S. Department of Energy, Washington, D.C., Report DOE/EIA-0384(92), 350 pp.

Enron Corporation, 1993, The outlook for natural gas: Houston, Tex., 16 pp.

Fisher, W. L., 1987, Can the U.S. oil and gas resource base support sustained production?: Science, v. 236, p. 1631-1636.

Fisher, W. L., 1988, The technological dependency of the remaining oil and gas resource base in the U.S.: Presentation to the Society of Petroleum Engineers, Symposium on Energy, Finance, and Taxation Policies, Washington, D.C.

Fisher, W. L., 1993, Statement to the U.S. Senate Committee on Energy and Natural Resources, Senator J. Bennett Johnston, Chairman, March 5, 1993, Houston, Tex., 11 pp.

Fletcher, S., 1992, Texas oilman: What about us?: The Houston Post, April 5.

Independent Petroleum Association of America (IPAA), 1991, Cost and Finance Study Committee Report.

Iseley, T., 1993, Personal communication.

Kaback, D. S., Looney, B. B., Corey, J. C., Wright, L. M., and Steele, J. L., 1989a, Horizontal Wells for In-Situ Remediation of Groundwater and Soils: Westinghouse Savannah River Co., report prepared under DOE contract no. DE-AC09-76SR00001, 16 pp.

Kaback, D. S., Looney, B. B., Corey, J. C., and Wright, L. M., 1989b, Well Completion Report on Installation of Horizontal Wells for In-Situ Remediation Tests: Westinghouse Savannah River Co., report prepared under DOE contract no. DE-AC090-88SR18035, 185 pp.

Kito, S., 1993, Analysis of the impact of the end of the fixed energy price period under interim Standard Offer 4 Contracts: Geothermal Resources Council Bulletin, v. 22 (3), p. 61-67.

Kramer, S. R., McDonald, W. J., and Thomson, J. C., 1992, An Introduction to Trenchless Technology: New York, Van Nostrand Reinhold, 223 pp.

MacGregor, I., 1993, National Science Foundation, Personal communication.

Muffler, L. J. P. (Editor), 1979, Assessment of Geothermal Resources of the United States—1978: U.S. Geological Survey Circular 790, 163 pp.

National Petroleum Council, 1992, The Potential for Natural Gas in the United States, Washington, D.C., NPC, 7 v.

National Research Council, 1979, Continental Scientific Drilling Program: Washington, D.C., National Academy Press, 192 pp.

National Research Council, 1988, Scientific Drilling and Hydrocarbon Resources: Washington, D.C., National Academy Press, 89 pp.

National Research Council, 1992, A Review of the Ocean Drilling Program Long Range Plan: Washington, D.C., National Academy Press, 13 pp.

Nuclear Waste News, 1991, v. 11 (49), p. 484, 12 Dec. 1991.

Nuclear Waste News, 1992a, v. 12 (14), p. 129, 2 April 1992.

Nuclear Waste News, 1992b, v. 12 (43), p. 404, 29 October 1992.

Nuclear Waste News, 1992c, v. 12 (15), p. 135, 9 April 1992.

Nuclear Waste News, 1992d, v. 12 (47), p. 439, 3 December 1992.

Nuclear Waste News, 1992e, v. 12 (40), p. 369, 8 October 1992.

Nuclear Waste News, 1992f, v. 12 (7), p. 57, 13 February 1992 (quoted by L. Duffy).

Oil Resources Panel, 1992, An assessment of the oil resource base of the United States, Commentary by Fisher, W. L., Tyler, N., Ruthven,

C. L., Burchfield, T. E., and Pautz, J. F.: Washington, D.C., U.S. Department of Energy, Report DOE/BC-93/1/SP, 54 pp.

Ozdimir, L., 1993, Colorado School of Mines, Personal communication.

Rex, R. W., and Howell, D. J., 1973, Assessment of U.S. geothermal resources: *in* Geothermal Energy, Kruger, P., and Otte, C., eds., Stanford University Press, p. 59-67.

Simon, R., and Cooper, G. A., 1994, Use of cryogenic fluids for environmental drilling in unconsolidated formations: *in* Drilling Technology, Vonzia, J. P., ed., Presentations at Energy-Sources Technology Conference and Exhibition, New Orleans, La., January 23-26, v. 56, p. 199-206.

U.S. Department of Energy, 1991, National Energy Strategy: Washington, D.C., U.S. Government Printing Office, 217 pp.

U.S. Department of Energy, 1992, Natural Gas Strategic Plan and Multi-Year Program Crosscut Plan—FY 1993-1998: Washington, D.C., Report DOE/FE-0251P, 231 pp.

U.S. Department of Energy, 1993, U.S. Geothermal Electric Power Plants—Data Tables: prepared by Meridian Coporation, Alexandria, Va.

U.S. Department of the Interior, 1989, Estimates of Undiscovered Conventional Oil and Gas Resources in the United States: A Part of the Nation's Energy Endowment: Washington, D.C., 44 pp.

Vincken, L. M. J., 1987, The future for drilling equipment and techniques: 12th World Petroleum Congress, Houston, Tex., April 26 - May 1, Production, v. 3, p. 119-130.

Volk, B. W., 1992, Results of Testing the Sonic Drilling System at the Hanford Site (September 1991 to May 1992): Report WHC-SD-EN-TRP-002, prepared for the U.S. Department of Energy by Westinghouse Hanford Company, Richland, Wash., 22 pp.

Volk, B. W., McLellan, G. W., Moak, D. J., Lerch, R. E., Thompson, K. M., and Barrow, J. C., 1993, ResonantSonic Drilling: History, Progress and Advances in Environmental Restoration Programs: Report WHC-SA-1959-FP, prepared for the U.S. Department of Energy by Westinghouse Hanford Company, Richland, Washington, 21 pp.

Westinghouse Savannah River Co., 1992, Demonstration of Eastman Christensen Horizontal Drilling System: Integrated Demonstration Site, Savannah River Site: Westinghouse Savannah River Co., report prepared under contract no. DE-AC09-89SR18035, 173 pp.

Woods, T. J., 1993, The Long-Term Trends in U.S. Gas Supply and Prices: The Edition of the GRI Baseline Projection of U.S. Energy Supply and Demand to 2010: Gas Research Insights, Gas Research Institute, March, 105 pp.

World Oil, 1993, 48th Annual International Outlook: v. 214 (8), p. 26.

Zoback, M. D., Apel, R., Baumgartner, J., Brudy, M., Emmerman, R., Engaser, B., Fuchs, K., Kessels, W., Rischmiller, H., Rummel, R., and Vernik, L., 1993a, Upper-crustal strength inferred from stress measurements up to 6 km depth in the KTB borehole: Nature, v. 365, p. 633-635.

Zoback, M. D., Hickman, S. H., Yormker, L. W., and Ellsworth, W. L., 1993b, Scientific drilling into the San Andreas Fault Zone: Preliminary proposal to the Continental Dynamics Program of the National Science Foundation.

4
DRILLING AND BORING OF ROCK

Introduction

In the smart drilling approach, whose development is recommended in this report, the process of local rock fracture, comminution, or other novel forms of rock removal is the first, and perhaps most crucial, step in a long succession of processes of reaching a target stratum or resource reservoir. As described in Chapter 5, mechanical fracturing of rock is still the most effective means of advancing the drill head. Thus, in the proposed ideal smart drilling approach, the drill must sense the type of rock or stratum ahead of the drill bit; recognize its resistance; and automatically adjust the drilling process in terms of rate, contact pressure, and so forth. If necessary, a smart drill will divert around a particularly difficult heterogeneity or seek alternative directions to avoid premature termination of the drilling operation.

Although identifying and evaluating rock ahead of the drill bit will require sophisticated advances in sensing and guidance, the actual fracturing, or comminution, of the rock requires a thorough understanding of the mechanical properties of rock and its response under the drill bits or cutters. Currently, there is insufficient understanding of both the interaction of cutting tools with the rock and the possible variations of such interactions based on rock type.

This chapter presents a summary of the current understanding of the structure and mechanical properties of rock, with particular emphasis on its fracture in compression, both under quasi-homogeneous stress fields and under conditions resembling the interaction of rock with the drill bit. The chapter is partly abstracted from a 1983 report by the National Materials Advisory Board (NMAB, 1983), but new developments have been added.

Types of Rock

The term "rock" includes a great variety of material types with distinctive characteristics. Rock is one of the materials for which fracture behavior under compressive stresses has been studied most thoroughly. For example, granitic rocks can behave in a brittle manner up to a confining pressure of 1 gigapascal (GPa), whereas carbonate rocks become plastic at moderate pressures of about 100 megapascals (MPa). Extensive crystal plasticity is observed in rock salt at moderate stress at room temperature, whereas most quartz-bearing rocks do not show significant dislocation activities up to about 400°C.

Rocks tend to be permeated with pores and microcavities, which were either formed during the inception of the rock or produced by its subsequent stress history. The porosity and microcavity morphology of rocks are as important as the mineralogic composition itself. Collectively, the microcavities cause nonlinear behaviors in many mechanical properties. These manifest themselves in the stress or pressure dependence of strain, velocity of sound, stress wave attenuation, and fracture behavior. Microcavities also introduce a scale effect into the prediction of mechanical behavior, and heterogeneities in the form of distribution of microcavities are the principal source of scatter in test results. Thus, this relatively readily probed characteristic of rocks may become an important indicator of mechanical properties to be sensed by the smart drill.

Rock has a finite hydraulic conductivity, which implies that a portion of the void space forms an interconnected network. Petrological and geophysical evidence indicates that rocks are saturated with water to a depth of tens of kilometers. Pore fluids play a significant role in engineering applications for energy resource recovery or dam construction. The effect of pore fluids on fracture behavior can be either mechanical through pore pressure diffusion or chemical through stress corrosion (NMAB, 1983). The effect of pore fluid will be an important measurable indicator of the mechanical properties of rock relevant to drilling.

Theoretical Models for the Behavior of Rocks in Compression

Like all solids, rocks can undergo true intrinsic inelastic behavior by dislocation motion, diffusional flow, or analogous processes occurring in glassy media. In most crustal rocks of interest, however, such intrinsic inelastic behavior is exhibited only at elevated temperatures or pressures that are not encountered during mechanical drilling. In most drilling applications, rocks act as purely elastic solids, but the heterogeneities discussed above can affect their elastic behavior. For equiaxed heterogeneities, rock behavior can be readily accounted for by a variety of bounding approaches, or more elegantly by ellipsoidal inclusion models (Chow, 1978) or self-consistent models (Budiansky, 1965). Corresponding approaches that account for the effects of microcracks are also well developed (Salganik, 1973; Budiansky and O'Connell, 1976). Overviews of such self-consistent models have been given by a number of authors (Cleary and others, 1980; Haskin, 1988; Kachanov, 1992). Thus, considerable theoretical and mechanistic methodologies exist that are capable of relating elastic properties to heterogeneities in the rock that govern its fracture behavior.

The apparent inelastic behavior of rocks known as *clastic flow* results from brittle fracture processes due to the formation and stable growth of brittle microcracks. This behavior has been dealt with in two ways. The first, a purely phenomenological approach, is used widely by civil engineers for characterizing the related clastic flow of concrete. Specific developments of this behavior by so-called deformation theories (Kupfer and Gerstle, 1973; Bazant and Tsubaki, 1980) or by incremental flow theories (Bazant and Kim, 1979; Nemat-Nasser and Shokovoh, 1980) have been discussed in the above-mentioned NMAB (1983) report. As discussed below, these formalisms are quite useful for predicting the development of shear faulting zones in rock, which is a central mechanism of rock fracture in drilling. They also find ready application in the understanding of chip formation under the drill bit.

Of fundamental interest to drilling is the behavior of fractured rocks in compression. Griffith (1924), who pioneered the understanding of brittle fracture in tension, also was the first to elucidate brittle fracture in compression. He noted that in a solid containing microcracks of many orientations under triaxial compression, local tensile fracture is initiated when shear stresses produced by unequal compression displace surfaces of

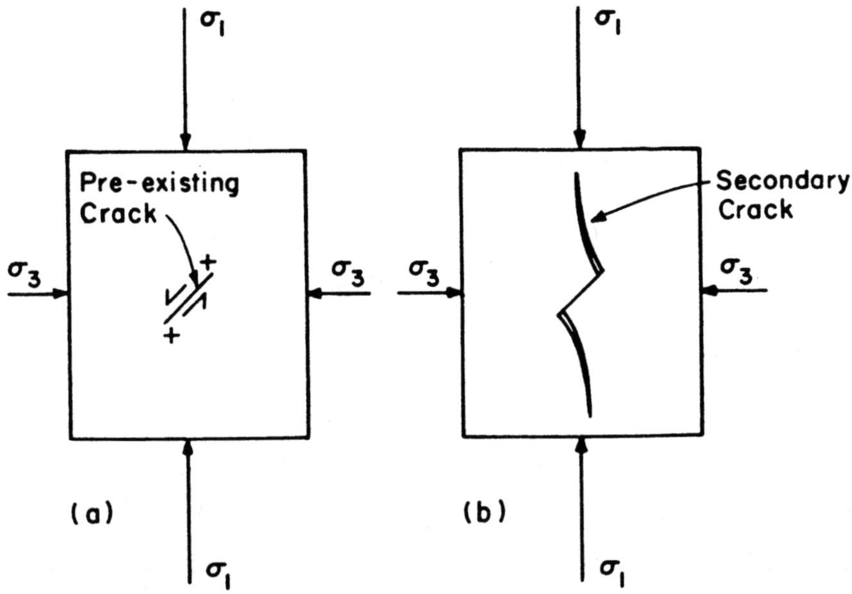

FIGURE 4.1 Extrinsic fracture in compression initiated from a large crack. (a) Stresses are concentrated at the ends of a shearing crack. (b) Microcracks extend from the shear crack parallel to the principal compression direction (NMAB, 1983).

preexisting microcracks. These displacements produce tensile stresses near the tips of these microcracks, which result in their growth across the local maximum tensile stress when this stress reaches the cohesive strength (Figure 4.1a). Griffith used this model of crack initiation to show that the uniaxial compressive strength of brittle solids must be many times (about eightfold) higher than their tensile strength, and that this compressive strength could be increased monotonically by the application of a confining pressure. It later became clear that the Griffith model needed modification to account for the frictional resistance on the touching faces of the microcracks; such a modified model was provided by McClintock and Walsh (1962). Figure 4.2 shows the predictions of the models of Griffith (1924) and McClintock and Walsh (1962) compared to experimental measurements (Ohnaka, 1973; NMAB, 1983). Clearly, the McClintock and Walsh model tends to track the experimental data better than the original Griffith model.

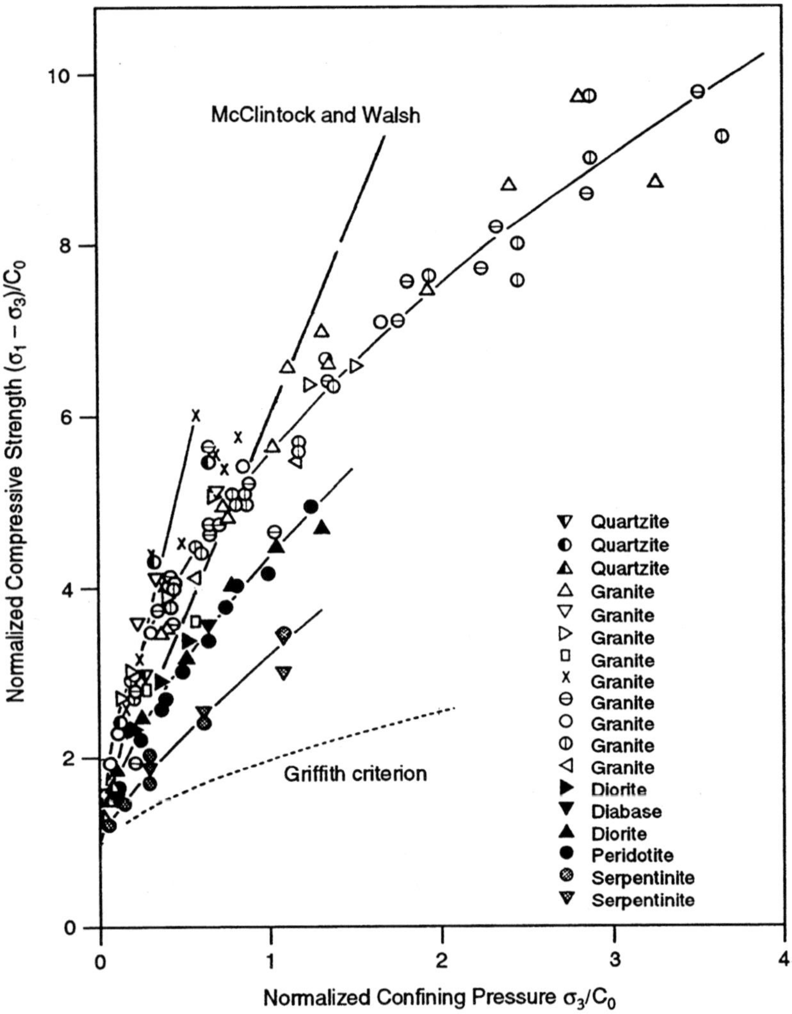

FIGURE 4.2 Relation between compressive strength and confining pressure for a series of rocks tested at room temperature (after Ohnaka, 1973).

It has been recognized by many investigators (e.g., Brace and Bombolakis, 1963; Erdogan and Sih, 1963; McClintock, 1965; Babel and Sines, 1968; Nemat-Nasser and Horii, 1982; Ashby and Hallam, 1986) that when the local recracking condition is satisfied, wing cracks develop from the extremities of the initial microcracks and extend stably in the direction of principal compression as shown in Figure 4.1b. Since the continued extension of the wing cracks necessitates continued relative translation of the faces of the initial microcracks, and since the total amount of such displacement is limited by their original lengths, overall compressive failure cannot result from the limited growth of wing cracks in "parts" (individual rock pieces or fragments) with dimensions much larger than those of the initial microcracks (NMAB, 1983). Thus, it is remarkable that the agreement between experiments and the simple recracking models is as good as that presented in Figure 4.2.

It is well known that brittle fracture in compression in massive parts occurs through the development of shear faults. This occurs when many developing wing cracks, aligned in a plane, begin to interact strongly en echelon and form a nucleus of a more macroscopic shear fault (Figures 4.3 a-b), which then spreads longitudinally to result in overall fracture. Although the conditions for development of the nucleus of the shear fault have not been well studied, the overall asymptotic conditions for their localization have been developed as a bifurcation model by Rudnicki and Rice (1975). They found that the critical strain at which the shear fault can develop freely depends strongly on the incremental moduli of the material containing the accumulating microcracks. The best agreement between theory and experiment, at least qualitatively, is obtained with a deformation theory approach. However, the actual conditions for fault development depend sensitively on the presence of initial imperfections, as is the case for all bifurcation phenomena (NMAB, 1983). The actual mechanism of brittle fracture in compression is a simple criterion of an equality between a crack driving force K_I and a material fracture toughness K_{Ic} at the tips of the microcracks, interacting en echelon. However, the development of a shear fault that produces eventual failure or chip formation in drilling obeys a phenomenological pressure-dependent deviatoric stress criterion (NMAB, 1983). Such mechanistic understanding of the development of shear faults should be most useful in the effective control of the smart drilling process.

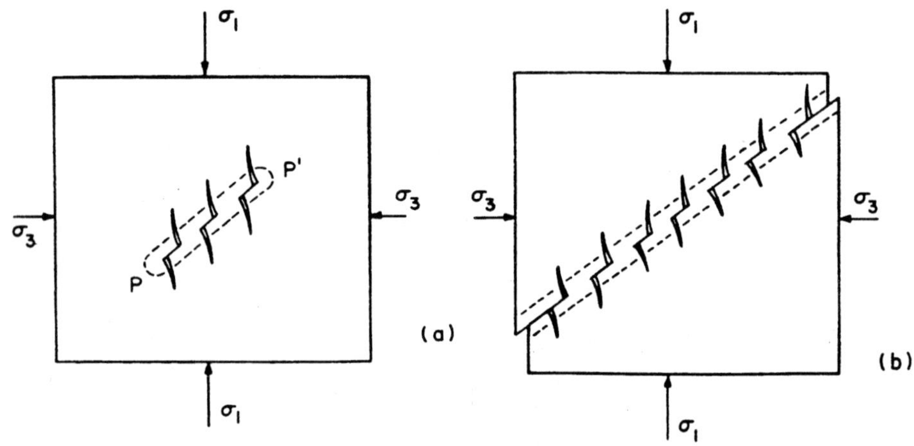

FIGURE 4.3 Development of intrinsic fracture behavior by initiation and propagation of a shear fault in compression. (a) Beginning of en echelon alignment of adjacent microcracks, producing a zone of increased shear compliance. (b) Idealized spread of the high-compliance zone across the part, resulting in a shear fault (NMAB, 1983).

Experimental Verification of Models

Two Views of Rock Behavior

Two approaches have been taken in the study of the mechanical behavior of rocks that together have contributed significantly to current understanding. The first is a global "applied mechanics" approach commonly used for engineering design and for the analysis of geologic faulting. The fracture process is taken as a discrete event without significant prior deformation and without warning. The only physical quantity of interest is the peak stress, which is of interest as an upper bound on solutions of the relevant boundary value problems. The fundamentals of this approach have been covered by many investigations including Jaeger and Cook (1979), Goodman (1980), Hoek and Brown

(1980), and Germanovich and Cherepanov (1987). A previous report by the U.S. National Committee for Rock Mechanics of the National Research Council (NRC, 1981) also touches on this subject. The second is a "mechanistic" approach that supplements standard deformation tests with nondestructive evaluation and microscopy, aiming at a fundamental understanding of the microscopic mechanism. The evolution of microstructure is treated as a continuous process culminating in the coalescence of microfissures to form a throughgoing fault. This approach was adopted by Paterson (1978) in his extensive review of rock fracture. It is this latter approach that has led to more definitive mechanistic understanding of the clastic behavior of rocks and their eventual fracture, which is discussed in a following section.

Laboratory Experiments on Clastic Flow and Fracture of Rocks

As noted previously, the true plastic response of rocks by dislocation motion occurs at such high temperature and pressure as to be of no relevance to the problem of mechanical rock drilling. Many constituent minerals in rocks can, of course, undergo twinning that, for all practical purposes, is not a thermally assisted rate process but usually requires very high stresses. Moreover, twinning is a very inhomogeneous form of deformation, and in the absence of local plasticity, it can at best influence only the fracture behavior of rocks. Thus, experimental studies carried out in the laboratory have been devoted largely to understanding the complex processes of microcracking under compression that result in clastic flow, which is a forerunner of the eventual shear faulting process discussed previously.

The clastic flow response of some common rocks such as Indiana limestone (Myer and others, 1992), Berea sandstone (Myer and others, 1992; Wong and others, 1992), and Carrara marble (Fredrich and others, 1989) has been investigated in the laboratory in a series of elegant experiments. These experiments have identified the key microstructural damage processes, which include fracturing of weak interface boundaries and grain boundaries; Hertzian diametral fracturing of grains; relative sliding across separated interfaces or boundaries, which act as inclined planes that "jack open" tensile wing cracks running parallel to the principal compression direction; and finally, en echelon action of interacting wing cracks that results in zones of shear faulting. In porous rock, pores may be crushed closed, which results in substantial permanent compression strains that locally stiffen and strengthen the rock.

In laboratory simulations of borehole breakout experiments on externally pressurized thick-wall cylinders of both Indiana limestone and Berea sandstone (Ewy and Cook, 1990a,b), the localization of en echelon damage has been followed in detail, leading to local breakout (see also Dyskin and Germanovich, 1993; Dyskin and others, 1993). These breakout experiments have demonstrated the anisotropic nature of such rocks and have also demonstrated a substantial size effect on rock strength. In particular, rocks in such breakout experiments exhibit a two- to threefold increase in local tangential splitting strength over laboratory-size cylindrical rock samples undergoing fracture by vertical spalling. This size effect, which clearly can be important in the actual fracture response of rock under drill bits, is not well understood and needs further study.

Compressive strength measurements of about 140 MPa, which have been obtained from the laboratory borehole breakout simulations, are quite high. Even when divided by 2.5, for example, to account for the size effect, these effective strengths are between 5 and 6 times higher than the often-quoted estimates of upper crustal effective strengths of about 10 MPa, which are commonly assumed to be proper for rock strength governed by frictional resistance (Zoback and others, 1993). Nevertheless, the strength levels measured in the laboratory may be more appropriate for understanding the local fracture strength of rock in the small volumes subjected to contact pressure by a drill bit. Clearly, these factors must be well understood to at least develop proper strength scaling relations that will be relevant in any definitive model of drilling.

Effect of Confining Pressure and Temperature on Strength

The brittle strength of rocks shows a strong pressure dependence because of its fractured state. It is not uncommon to achieve a tenfold increase in strength by a moderate increment in mean confining stress, as shown in Figure 4.2. Depending on the signs of the principal stresses, a sample can fail in tension or in shear. However, only shear fracture that occurs when the principal stresses are all compressive (Figure 4.4) is considered here. The fracture angle is defined as the angle between the shear fault surface and the maximum principal compressive stress σ_1.

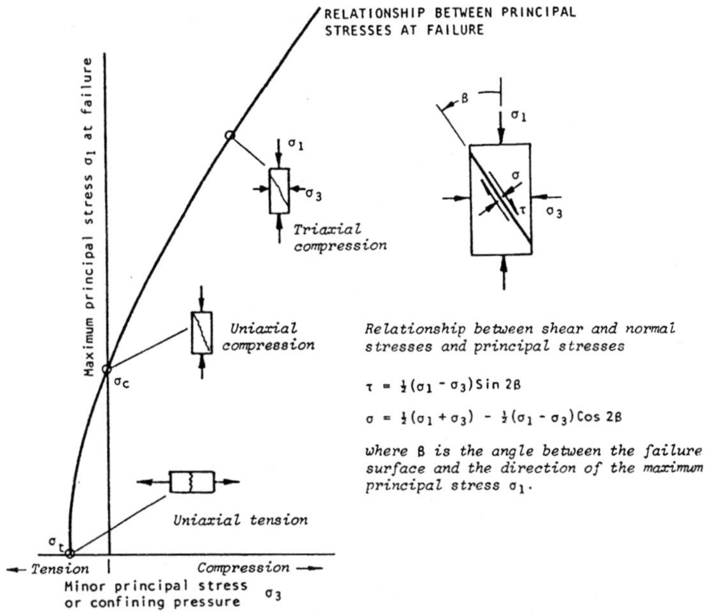

FIGURE 4.4 Relationship between shear and normal stresses and principal stresses at failure ranging from tensile, to uniaxial compression, to compression under a confining pressure (Hoek and Brown, 1980).

Most of the empirical fracture criteria discussed earlier are formulated from conventional triaxial test data with the implicit assumption of the independence of the fracture phenomenon on the intermediate principal stress σ_2. However, Mogi (1972) concluded from his "true" triaxial tests on cube-shaped samples that this is only an approximation. He found that when the minimum compressive stress σ_3 is kept constant, an increase in σ_2 results in an increase of σ_1 at failure (Figure 4.5a) and, furthermore, that the fracture angle decreases with increasing σ_2 when σ_3 is fixed (Figure 4.5b). The fracture surface, however, always contains the σ_2 direction. Ductility also is found to decrease with an increase in σ_2. Similar conclusions were drawn by Handin and others (1960, 1963) from torsion tests on hollow cylindrical specimens.

The effect of temperature increases is to stabilize postfailure behavior (Wong, 1982). In general, in comparison with pressure, temperature has a relatively small effect on the brittle fracture of dry rocks.

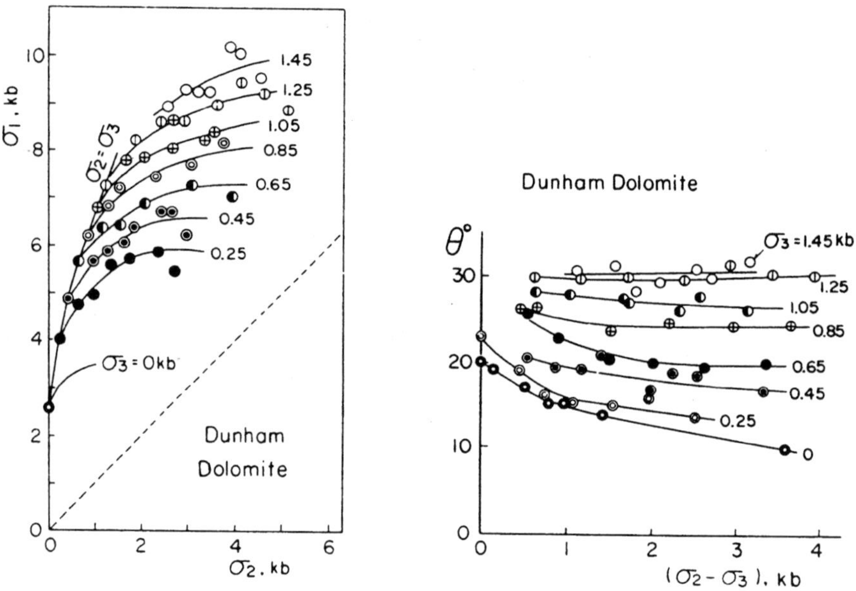

FIGURE 4.5 (a) Dependence of compressive strength σ_1 on transverse compressive stress σ_2 for different levels of the third principal compressive stress σ_3 from tests in triaxial compression. (b) Dependence of the fracture angle θ, between the shear fault plane and the direction of principal compression, on the principal stress difference from tests in triaxial compression (Mogi, 1972).

Effect of Pore Fluid

Experiments on sedimentary rocks by Handin and others (1963) and on crystalline rocks by Brace and Martin (1968) show that if the sample is "drained," Terzaghi's principle of effective stress should apply to fracture (Jaeger and Cook 1979). If pore fluid diffusion is relatively slow (so that pore pressure is no longer uniform), it is necessary to take into account microcavity deformation and permeability. Theories for such so-called poroelastic behavior have been developed primarily in petroleum engineering (Biot, 1941) and have been reviewed by Rice and Cleary (1976; see also Atkinson and Craster, 1991; Detournay and Cheng, 1991). Clearly, these developments will be of importance in tailoring the drilling process to local requirements.

If the interstitial fluid is not inert relative to the mineral constituents of the rock, the pore fluid can exert a chemical effect in addition to the purely mechanical one discussed above. An evident weakening effect in water-saturated samples has been observed in calcite by Rutter (1972) and in quartz by Scholz (1968c). The observed behavior is usually attributed to stress corrosion cracking (Atkinson, 1982). Mizutani and others (1977) investigated the strengthening effect of a sample when placed in a high vacuum similar to the lunar environment. A limited amount of available data indicates that the stress corrosion cracking effect is reduced by an increase in pressure (Kranz, 1980) or a decrease in temperature (Kranz and others, 1982).

Effect of Size

Because of their brittle nature, the compressive strength of rocks depends on size, as pointed out above (see also NMAB, 1983). Size effects are particularly important in relating laboratory fracture experiments both to failures in massive crustal formations and to the high-gradient local fracture processes under the drill bit.

Evolution of Sources of Shear Faulting

The sequence of processes of microcrack interactions leading to the evolution of shear faults has been studied in the laboratory both by techniques of systematic sectioning experiments and by a variety of acoustic techniques. For example, monitoring the acoustic emission (AE) from the compressed samples by multiple probes has permitted determination of the spatial correlation of signals emanating from microcracking events leading to formation of shear faults. Such sophisticated AE studies (Mogi, 1968, 1972; Scholz, 1968a, b; Byerlee and Lockner, 1977; Lockner and others, 1977; Rothman, 1977; Bailey and others, 1979; Lockner and Byerlee, 1980; Sondergeld and Estey, 1981) support the microstructural studies discussed previously and show that existing imperfections are likely sites of AE activity. Microcracking is found to intensify in the neighborhood of previous microcrack sites and is often persistent in a locality until final shear faulting sets in (Sondergeld and Estey, 1982).

Recent experimental observations on the evolution of microcracking damage in rocks and scanning electron microscopy (SEM) experiments provide graphic means for understanding shear localization measurements by AE. Central features in the development of a faulting nucleus are the en echelon interaction of microcracks and the planar form of dilatancy they introduce into the fracture problem. Tapponier and Brace (1976) concluded that dilatancy is primarily a consequence of two types of cracking: (1) widening and extension of preexisting discontinuities such as grain boundaries, cracks, and pores; and (2) initiation and propagation of cracks at heterogeneity sites. These observations clearly demonstrate that a continuum description of the fracture localization behavior of rocks is adequate only over a so-called representative continuum volume element large enough to smooth out grain-scale inhomogeneities. This presents a problem of how to translate laboratory findings in quasi-homogeneous deformation fields to the high-gradient fields under a drill bit or cutter, for eventual application for control of cutting operations by a smart drill.

The SEM observations show, for example, that quartz, which comprises about one-third by volume of granite, has limited participation in the localization process in the initial postfailure stage. In other words, localized deformation extending over a continuum element with grains of all major mineral types is not observed until the sample has been deformed well into the postfailure stage. In this limiting sense, SEM observations agree with the theoretical prediction of localization analysis outlined previously for frictional, dilatant materials. Such analyses (e.g., Rudnicki and Rice, 1975; Rudnicki, 1977) predict that the onset of localization under axisymmetric compression should occur when the sample has been deformed well into the strain-softening stage—another factor of importance in the mechanistic rationalization of the drilling process.

Surface-Active Agents in Rock Fracture

In brittle fracture of solids, the work of fracture results largely from the surface free energy of the solid (Griffith, 1920). Even in pseudoductile fracture where the actual fracture process is one of extension of brittle cracks, the surface free energy is the overall factor that scales to specific fracture work, even though other energy-dissipating phenomena such as frictional rubbing and true plastic deformation are present (Rice, 1965). Under these conditions, the presence of surface-active agents, which significantly reduce the free energy of the surfaces that have been created by fracture, can have dramatic effects on the overall specific work of fracture, known collectively as Rehbinder effects (Rehbinder and Shchukin, 1972).

There are many examples of dramatic reduction of the local work of brittle separation by active liquid environments in laboratory experiments (Westwood and Pickens, 1983). Although the fundamental mechanisms of these effects are not fully understood, they are often well characterized, and some have been used with mixed success in near-surface drilling (Zoback and others, 1993). Some evidence (Kranz, 1980) suggests, however, that such Rehbinder effects are considerably reduced when the part is under pressure or when the surface-active agent cannot penetrate to the crack tips, particularly when the key shear faulting events are of a subsurface nature. Consequently, it is not clear whether these effects are present in deep drilling environments. Surface-active agents can also promote stress corrosion cracking of drilling equipment. However, this problem could probably be avoided by careful process planning.

Another important potential application of surface-active agents is their use to influence wear rates of diamond tool bits in cutters (Cooper and Berlie, 1978; Mills and Westwood, 1978). In a detailed experimental study, Cooper and coworkers (Cooper and Berlie, 1978; Cooper, 1979) have established that few, if any, surface-active agents have produced unambiguous weakening of a variety of rocks, including marble and granite. They have shown that these agents can either increase or decrease wear rates of the cutters, and that oxidizing agents in particular promote increased diamond wear. The best understanding of this effect is that these agents promote effective nucleate boiling heat transfer between the hot diamond and the cooling fluids, and that this appears to have a major beneficial effect in most cases.

Whatever the mechanism of the effect of the environment on the rock destruction process, be it by changing the strength of the rock or affecting the wear process of the tool, the effects are sufficiently dramatic, when present, that they merit further special consideration to exploit their full potential.

Fracturing of Rock in Drilling

Phenomenology of Drill-Rock Interaction

One of the important practical applications of fracture of brittle solids in compression is the penetration of drilling tools. Examples include percussive and rotary drilling, drag bit drilling, and ploughing or planing of rock formations (e.g., coal). Despite the extreme commercial importance of tool penetration into brittle rock, details of the penetration mechanism are poorly understood. Much of the past work on this problem has been directed toward semiempirical drilling models specific to a particular formation. Although such models have some engineering utility, they generally are not consistent with recent developments concerning the constitutive behavior of rock, and they are not based on the mechanistic processes described above. Hence, their range of applicability is quite limited. The following sections review the current practice in phenomenological theories for tool penetration into rock and the attempts at developing mechanistic models for chip formation. Quite clearly, a detailed understanding of these will be of key importance in the establishment of controlled cutting strategies for exploitation of the full potential of the smart drilling process.

Quasi-static tool penetration tests (Evans and Murrell, 1958; Hartman, 1959; Reichmuth, 1963; Sikarskie, 1966; Singh and Johnson, 1967) have established the features of the penetration mechanism. Figure 4.6 shows a typical set of force-penetration traces of wedge-shaped tools into charcoal gray granite. As the traces suggest, the penetration process occurs in a repetition of two distinct phases. The first is a crushing phase, in which the forces on the tool increase monotonically. Hydrostatic stresses in the vicinity of the tool tip are extremely high; the material under the tip is crushed in the region of very high contact pressures and undergoes a volumetric expansion. The crushed and expanded material behaves in an almost plastic fashion. The tool-bit stresses are transmitted through the crushed zone, resulting in a stress field closely resembling that of a plastic indentation problem. In the second phase, a macrofracture zone is initiated, and with a subsequent load increase, it eventually grows to form a chip. The cycles repeat under increasing total force as the contact area between the tool and the rock increases.

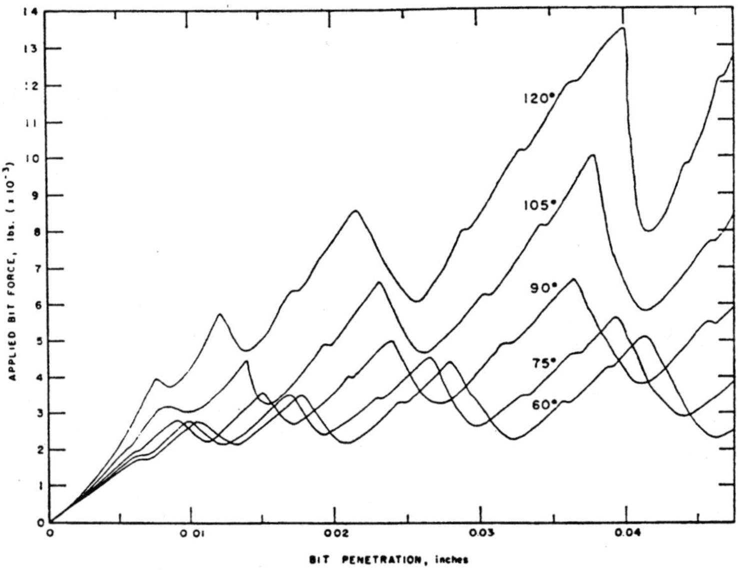

FIGURE 4.6 Characteristic force-penetration curves for charcoal gray granite by 2.5-cm-wide wedges (Reichmuth, 1963).

A typical drill bit action may see two to four cycles per "blow." The cyclic behavior is also observed in drag bits (Whittaker and Szwilskie, 1973). This cyclic penetration behavior applies to rock at low confining pressure. At high confining pressure (e.g., in deep drilling), rock develops more prominent clastic flow behavior and the force-penetration curves become smoother (Cheatham and Sikarskie, 1973).

Theoretical "Plasticity" Models for the Faulting of Rocks

The process of chip formation as outlined above involves the repeated application of local shear faulting in the rock under the concentrated pressure of the tool. The mechanics and mechanisms of the evolution of microcracking processes leading to the formation of a shear fault nucleus in a homogeneous compressive stress field under confining pressure apply locally in the rock that is to be chipped—albeit in this case, the local stress field is highly inhomogeneous, and shear faults comprise curved surfaces.

In such situations, more formal phenomenological theories clearly are needed to deal with the problem on the basis of a continuum, without repeatedly coming to grips with the details of faulting. The constitutive localization model of Rudnicki and Rice (1975), which develops conditions of shear localization by treating the brittle rock undergoing microcracking as if it were a pressure-sensitive, somewhat dilating plastic continuum, serves as a guide. These problems may be amenable to analysis using the slip line field method of the mathematical theory of plane plasticity, generalized to deal with the pressure dependence of plastic resistance and dilatancy, and including, if necessary, vertex phenomena on yield surfaces that are known to make the material more prone to localization of deformation. This approach is currently under active development (Mokhel and Salganik, 1993). A number of attempts have been made to model the behavior of granular materials such as soils (Drucker and Prager, 1952; Drescher and others, 1967; Szczepinski, 1971; Mroz and Szymanski, 1978; Spencer, 1982; Anand, 1983). These models have been applied with varying degrees of success to inhomogeneous problems, such as indentation of a half space of sand or soil, but they have not found application to corresponding problems of machining or chip formation in rocks where modifications that take into account the small size of the stressed volume, the scale of heterogeneities, and the effects of pore fluids become necessary. This will be a fertile area for immediate development for application to the smart drilling process.

Specific Models for Chip Formation

Theoretical models using slip line plasticity approaches have been developed specifically for the purpose of dealing with chip formation in rocks by the wedge indentation process. Figure 4.7 shows a schematic two-dimensional view of a wedge penetration model developed by Paul and Sikarskie (1965). The plane strain model has assumed symmetrical distortions and an initial local penetration by crushing that is linear with the applied force. The model further assumes that the chips are planar blocks and that they move out of the way when the Coulomb-Mohr criterion of a pressure-dependent plastic resistance can be counteracted. The model demonstrates that for proper chip formation, the wedge angle θ shown in Figure 4.7 must be less than a given amount, determined by the

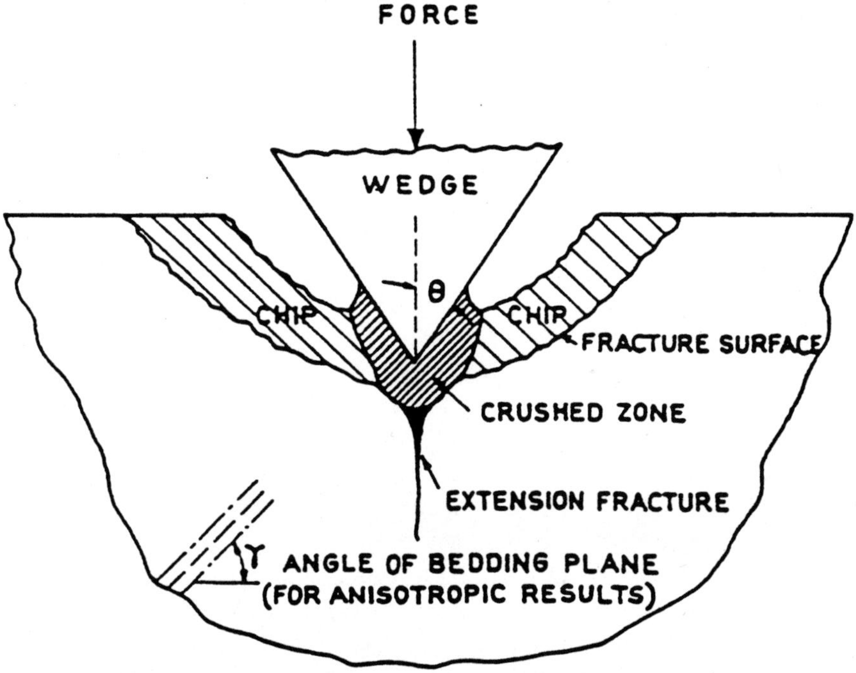

FIGURE 4.7 Idealized model of the penetration of a tool wedge into rock (NMAB, 1983).

friction angle between the wedge and the rock, and the pressure dependence of the plastic resistance that governs the Coulomb-Mohr criterion. The model is capable of presenting an upper-bound, outer envelope to the force-penetration behavior of a wedge as given in Figure 4.6. The model has been extended to the symmetrical penetration of anisotropic rock by Benjumea and Sikarskie (1969) and to the application of a tilted wedge by McLamore (1971).

The simple wedge penetration model described above is not fully consistent with the actual phenomenon. Altiero and Sikarskie (1974) have studied this in some detail on models made of plaster of Paris to better understand the complex processes of vertical splitting, crushing, and chip formation. On the basis of their observations, Sikarskie and Altiero (1973) analyzed the wedge penetration process as a quarter-space loading problem, where the free surface and the vertical splitting delineate the quarter-space boundaries as shown in Figure 4.7. The normal and frictional loading of the interface between the rock and the wedge is then considered in detail, the stress distribution in the rock is calculated, and the Coulomb-Mohr criterion is applied to determine the point at which fracturing will begin. The growth direction of the chip was assumed to follow the steepest slope of the Coulomb-Mohr function. This is clearly somewhat arbitrary; the generalized slip line field approach discussed above probably should have been used instead.

Other Applications of Rock Fracture

Other applications of rock fracturing are of industrial interest. Hydraulic fracturing is of great interest in the improvement of oil recovery from oil-bearing rock beds. Here the problem is one of tensile fracturing of rock under a large wellbore pressure and the opposing confining pressures of the surrounding rock, made complicated by the flow processes of pore fluids. Problems of mine shaft failures and lateral displacement of large earthworks such as dams and embankments are direct applications of compressive fracturing of rock and soil flow, both obeying the general framework of shear faulting theories. The framework for the solution to such problems exists, and solutions will be urgently required in the context of the development of the smart drilling process.

Priorities for R&D

The rate-controlling process in drilling is, and will likely be for the foreseeable future, the rate of rock removal by the drill bit. Increasing the efficiency of the rock removal rate requires a fundamental understanding of how solid rock is fragmented by drill bits and the numerous factors that control this very inhomogeneous and local process. There are several areas in which additional research would be most beneficial in improving rock removal, particularly in the context of the *smart drilling system* elucidated in Chapter 2. These are broadly the following:

1. *In situ characterization of rock:* The chemical constitution (including mineralogy and petrology), physical characteristics, microstructure, and nature and distribution of preexisting flaws (e.g., faults, joints, and bedding planes) of rock have been studied in great detail for many decades. There have been equally detailed laboratory studies of the mechanical properties of rock over a wide range of temperature and pressure to identify mechanisms for drilling and excavation. In any new thrust to capitalize on the potential of the *smart drilling approach*, it is essential to tailor the response of the drill to the local rock environment (e.g., temperature, pressure, fracture density, and resistance). This could be achieved if the relevant mechanical properties of the rock could be assessed in situ by making meaningful measurements of its chemical, mineralogical, elastic, and acoustic properties, and if these measurements could be used to make adjustments to the drill bit in order to optimize the rate of rock removal.

2. *Fracture processes:* For drilling applications, the relevant rock behavior is purely elastic, with brittle fracture occurring under the drill bit. However, conditions under the drill bit differ significantly from the quasi-homogeneous conditions in the Earth's crust or in typical laboratory experiments. The stress field under the drill bit is highly inhomogeneous over volume elements much smaller than those encountered in the usual laboratory experiments, where rock fracture is subject to the strength-size effects discussed previously. Presently, there is inadequate information on the nature of this inhomogeneous fracture process, both qualitatively and quantitatively. The fracture process should be studied in greater detail to identify the specific factors affecting both the local driving forces of fracture and the response of the rock to these forces. These processes should be modeled to obtain useful scaling laws for drilling practice.

3. *Matching the drilling process to the physical environment:* The requirements for effective drilling vary for different environments, for example, near-surface, large-scale excavation; porous and fluid-bearing rock; dense rock in the upper crust; and fractured, hot rock in geothermal reservoirs. The physical conditions and mechanical properties of rocks are well known in these environments, but there is inadequate information on how to optimize the drilling process to best suit these different requirements.

4. *Physicochemical effects on rock fracture and tool wear:* A potentially important avenue in the improvement of drilling practice involves the physicochemical aspects of the mechanical response of rock in surface-active fluid environments, to take advantage of the so-called Rehbinder effects. Every brittle fracture process is ultimately based on the severance of chemical bonds. There are many examples of dramatic reduction of the local work of brittle separation in active liquid environments. Although the mechanistic details of these effects are not fully understood, they are often well characterized, and some have been tried with some success in drilling. Because of their attractive potential, these effects should be given special consideration.

5. *Unconventional methods for rock removal:* Rock is a brittle solid with a characteristic set of strength-impairing imperfections such as pores, weak boundaries, and microcracks. The tensile strength of unconfined rock is generally low, approaching at best 50 to 100 MPa. When confined, or under pressure, as is the case at great depths in the Earth's crust, these imperfections are rendered ineffective by being pressed shut. Under these conditions, the compressive strength of rock is high and increases with increasing pressure. Thus, drilling rock at great depth by subjecting it locally to greater pressure is probably an inefficient means of removing material. Efforts should be made to identify unconventional forms of material removal that capitalize on the low tensile strength of rocks.

A large number of unconventional methods of rock removal have been studied. When compared with mechanical drilling, these techniques have so far been found to be quite ineffective, as discussed in more detail in the next chapter. Nevertheless, the existing information on them should be reevaluated to reach firm quantitative conclusions concerning their potential, particularly for use in conjunction with mechanical drilling.

References

Altiero, N. J., and Sikarskie, D. L., 1974, Fracture initiation and propagation in elastic-brittle materials subjected to compressive stress fields—An experimental study: Mechanical Research Communication, v. 1 (4), p. 225-231.

Anand, L., 1983, Plane deformations of ideal granular materials: Journal of the Mechanics and Physics of Solids, v. 31 (2), p. 105-122.

Ashby, M. F., and Hallam, S. D., 1986, The failure of brittle solids containing small cracks under compressive stress states: Acta Metallurgica, v. 34 (3), p. 497-510.

Atkinson, B. K., 1982, Subcritical crack propagation in rocks: theory, experimental results and applications: Journal of Structural Geology, v. 4 (1), p. 41-56.

Atkinson, C., and Craster, R. V., 1991, Plane strain fracture in poroelastic media: Proceedings of the Royal Society, v. A434, p. 605-663.

Babel, H. W., and Sines, G., 1968, A biaxial fracture criterion for porous brittle materials: ASME, Journal of Basic Engineering, ser. D, v. 90 (2), p. 285-291.

Bailey, C. D., Hamilton, J. M., Jr., and Pless, W. M., 1979, Acoustic emission of impact-damaged graphite-epoxy composites: Materials Evaluation, v. 37 (6), p. 43-48.

Bazant, Z. P., and Kim, S. S., 1979, Plastic fracturing theory for concrete: Journal of Engineering Mechanics, v. 105 (3), p. 407-428.

Bazant, Z. P., and Tsubaki, T., 1980, Total strain theory and path-dependence of concrete: Journal of Engineering Mechanics, v. 106 (6), p. 1151-1173.

Benjumea, R., and Sikarskie, D. L., 1969, A note on the penetration of a rigid wedge into a nonisotropic brittle material: International Journal of Rock Mechanics and Mineral Science, v. 6 (4), p. 343-352.

Biot, M. A., 1941, General theory of three-dimensional consolidation: Journal of Applied Physics, v. 12 (2), p. 155-164.

Brace, W. F., and Bombolakis, E. G., 1963, A note on brittle crack growth in compression: Journal of Geophysical Research, v. 68 (12), p. 3709-3713.

Brace, W. F., and Martin, R. J., III, 1968, A test of the law of effective stress for crystalline rocks of low porosity: International Journal of Rock Mechanics and Mineral Science, v. 5 (5), p. 415-426.

Budiansky, B., 1965, On the elastic moduli of some heterogeneous materials: Journal Mechanics and Physics of Solids, v. 13 (4), p. 223-227.

Budiansky, B., and O'Connell, R. J., 1976, Elastic moduli of dry and saturated cracked solids: International Journal of Solids Structures, v. 12 (1), p. 81-97.

Byerlee, J. D., and Lockner, D., 1977, Acoustic emission during fluid injection into rock: Proceedings of the First Conference on Acoustic Emission/Microseismic Activity in Geological Structures and Materials, Handy, H. R., Jr., and Leighton, F. W., eds. Clausthal, FRG, Trans. Tech., p. 87-98.

Cheatham, J. B., and Sikarskie, D. L., 1973, Penetration problems in rock mechanics: Annual ASME Applied Mechanics Division of Rock Mechanics Winter Symposium, Sikarskie, D. L., ed., New York, American Society of Mechanical Engineers, v. 3, p.41-71.

Chow, T. S., 1978, Effect of particle shape at finite concentration on the elastic moduli of filled polymers: Journal of Polymer Science (Polymer Physics), v. 16 (6), p. 959-970.

Cleary, M. P., Chen, I.-W., and Lee, S.-M., 1980, Self consistent techniques for heterogeneous media: Journal of Engineering Mechanics, v. 106 (5), p. 861-887.

Cooper, G. A., and Berlie, J., 1978, Comments on influence of chemomechanically active fluids on diamond wear driving hard rock drilling: Journal of Materials Science, v. 13 (12), p. 2716-2723.

Cooper, G. A., 1979, Some observations on environmental effects when diamond drilling: *in* The Science of Ceramic Machining and Surface Finishing II, Hockey, B. J., and Rice, W. R., eds., NBS Special Publication 562, Washington, D.C., U.S. Government Printing Office, p. 115-138.

Detournay, E., and Cheng, A. H.-D., 1991, Plane strain analysis of a stationary hydraulic fracture in a poroelastic medium: International Journal of Solids and Structures, v. 27 (13), p. 1645-1662.

Drescher, A., Kwaszczynska, K., and Mroz, Z., 1967, Statics and kinematics of granular medium in case of wedge indentation: Archiwum Mechaniki Stosowanej, v. 19, p. 99.

Drucker, D. C., and Prager, W., 1952, Soil mechanics and plastic analysis or limit design: Quarterly of Applied Mathematics, v. 10, p. 157-166.

Dyskin, A. V., and Germanovich, L. N., 1993, Model of rock burst caused by cracks growing near free surface: Rockbursts and Seismicity in Mines '93, Young, P., ed., Rotterdam, Balkema, p. 169-175.

Dyskin, A. V., Germanovich, L. N., and Ustinov, K. B., 1993, A 2-D model of skin rock burst and its application to rock burst monitoring: *in* Geotechnical Instrumentation in Open Pit and Underground Mining, Szwedzicki, T., ed., Rottterdam, Balkema, p. 133-141.

Erdogan, F., and Sih, G. S., 1963, On the crack extension in plates under plan loading and transverse shear: ASME Journal of Basic Engineering, ser. D, v. 85 (4), p. 519-527.

Evans, I., and Murrell, S. A. F., 1958, Mechanical Properties of Nonmetallic Brittle Materials: New York, Interscience, 432 pp.

Ewy, R. T., and Cook, N. G. W., 1990a, Deformation and fracture around cylindrical openings in rock—I. Observations and analysis of deformations: International Journal of Rock Mechanics, Mineral Science, and Geomechanical Abstracts, v. 27 (5), p. 387-407.

Ewy, R. T., and Cook, N. G. W., 1990b, Deformation and fracture around cylindrical openings in rock—II. Initiation, growth and interaction of fractures: International Journal of Rock Mechanics, Mineral Science, and Geomechanical Abstracts, v. 27 (5), p. 409-427.

Fredrich, J. T., Evans, B., and Wong, T.-F., 1989, Micromechanics of the brittle to plastic transition in Carrara marble: Journal of Geophysical Research, v. 94 (B4), p. 4129-4145.

Germanovich, L. N., and Cherepanov, G. P., 1987, Fracture criteria for material with defects: Applied Mathematics and Mechanics, v. 51, p. 256-264.

Goodman, R. E., 1980, Introduction to Rock Mechanics: New York, J. Wiley and Sons, 475 pp.

Griffith, A. A., 1920, The phenomena of rupture and flow in solids: Philosophical Transactions of the Royal Society, v. A221, p. 163-198.

Griffith, A. A., 1924, Theory of rupture: Proceedings of the 1st International Conference on Applied Mechanics, Delft, Holland, p. 55-63.

Handin, J., Higgs, D. V., and O'Brien, J. K., 1960, Torsion of Yule marble under confining pressure: *in* Rock Deformation, Griggs, D. T., and Handin, J., eds., GSA Memoir No. 79, Boulder, Colo., Geological Society of America, p. 245-274.

Handin, J., Hager, V., Friedman, M., and Feather, J. N., 1963, Experimental deformation of sedimentary rocks under confining pressure: pore pressure tests: Bulletin of the American Association of Petroleum Geologists, v. 47 (5), p. 717-751.

Hartman, H. L., 1959, Basic studies of petroleum drilling: Transactions AIME, v. 24 (1), p. 68.

Haskin, Z., 1988, The differential scheme and its application to cracked materials: Journal of the Mechanics and Physics of Solids, v. 36 (6), p. 719-734.

Hoek, E., and Brown, E. T., 1980, Underground Excavations in Rock: London, Institute of Mining and Metallurgy, 138 pp.

Jaeger, J. C., and Cook, N. G. W., 1979, Fundamentals of Rock Mechanics: New York, Halsted Press, 593 pp.

Kachanov, M., 1992, Effective elastic properties of cracked solids: critical review of some basic concepts: Applied Mechanics Review, v. 45, (8), p. 304-335.

Kranz, R. L., 1980, The effects of confining pressure and stress difference on static fatigue of granite: Journal of Geophysical Research, v. 85 (B4), p. 1854-1866.

Kranz, R. L., Harns, W. J., and Carter, N. L., 1982, Static fatigue of granite at 200 °C: Geophysical Research Letters, v. 9 (1), p. 1-4.

Kupfer, H. B., and Gerstle, K. H., 1973, Behavior of concrete under biaxial stresses: Journal Engineering Mechanics, v. 99 (EM4), p. 853-866.

Lockner, D. A., Walsh, J. B., and Byerlee, J. D., 1977, Changes in seismic velocity and attenuation during deformation of granite: Journal of Geophysical Research, v. 82 (35), p. 5374-5378.

Lockner, D. A., and Byerlee, J. B., 1980, Development of fracture planes during creep of granite: Proceedings of the Second Conference on Acoustic Emission/Microseismic Activity in Geological Structures and Materials, Handy, H. R., Jr., and Leighton, F. W., eds., Clausthal, FRG, Trans. Tech., p. 11-25.

McClintock, F. A., 1965, Effects on root radius stress crack growth and rate on fracture instability: Proceedings of the Royal Society, v. A285, p. 58-72.

McClintock, F. A., and Walsh, J. B., 1962, Friction on Griffith cracks in rocks under pressure: Proceedings of the 4th U.S. National Congress of Applied Mechanics, v. 2, p. 1015.

McLamore, R. T., 1971, The role of rock strength anisotropy in natural hole deviation: Journal of Petroleum Technology, v. 23 (11), p. 1313-1321.

Mills, J. J., and Westwood, A. R. C., 1978, Influence of chemomechanically active fluids on diamond wear during hard rock drilling: Journal of Materials Science, v. 13 (12), p. 2713-2716.

Mizutani, H., Spetzler, H., Getting, I., Martin, R. J., III, and Soga, N., 1977, The effect of outgassing upon the closure of cracks and the strength of lunar analogues: Proceedings of the 8th Lunar Science Conference, New York, Pergamon Press, p. 1235-1248.

Mogi, K., 1968, Source location of elastic shocks in the fracturing process in rocks: Earthquake Research Institute, Bulletin, v. 46, p. 1103-1125.

Mogi, K., 1972, Fracture and flow of rocks: Tectonophysics, v. 13, p. 541-68.

Mokhel, A. N., and Salganik, R. L., 1994, Development of slip concept in the theory of plastic of materials and rocks: Journal of the Mechanical Behavior of Materials (in press).

Mroz, Z., and Szymanski, Cz., 1978, Non-associated flow rules in description of plastic flow of granular materials: *in* Limit Analysis and Rheological Approach in Soil Mechanics, Olszak, W., and Suklje, L., eds., New York, Springer, p. 49-94.

Myer, L. R., Kemeny, J. M., Zheng, Z., Suarez, R., Ewy, R. T., and Cook, N. G. W., 1992, Extensile cracking in porous rock under differential compressive stress: Applied Mechanics Review, v. 45 (8), p. 263.

National Materials Advisory Board, 1983, Fracture in Compression of Brittle Solids: NMAB-404, Washington, D.C., National Academy Press, 72 pp.

National Research Council, 1981, Rock-Mechanics Research Requirements for Resource Recovery, Construction, and Earthquake-Hazard Reduction: Washington, D.C., National Academy Press, 222 pp.

Nemat-Nasser, S., and Shokovoh, A., 1980, On finite plastic flows of compressible materials with internal friction: International Journal of Solids and Structures, v. 16 (6), p. 495-514.

Nemat-Nasser, S., and Horii, H., 1982, Compression-induced nonplanar crack extension with application to splitting, exfoliation, and rock burst: Journal of Geophysical Research, v. 87 (B8), p. 6805-6821.

Ohnaka, M., 1973, The qunatitative effect of hydrostatic pressure on the compressive strength of crystalline rods: Journal of the Physics of the Earth v. 21 (2), p. 125-140.

Paterson, M. S., 1978, Experimental Rock Deformation—The Brittle Field: New York, Springer, 254 pp.

Paul, B., and Sikarskie, D. L., 1965, A preliminary theory of static penetration by a rigid wedge into a brittle material: Transactions AIME, v. 232, p. 372.

Rehbinder, P. A., and Shchukin, E. D., 1972, Surface phenomena in solids during deformation and fracture: Progress in Surface Science, v. 3 (2), p. 97-188.

Reichmuth, D. R., 1963, Correlation of force-displacement data with physical properties of rock for percussive drilling systems: *in* Rock Mechanics, Fairhurst, C., ed., Oxford, Pergamon Press, p. 33-60.

Rice, J. R., 1965, An examination of the fracture mechanics energy balance from the point of view of continuum mechanics: Proceedings of the 1st International Conference on Fracture, Yokobori, T., Kawasaki, T., and Swadlow, J. L., eds., Sendai, Japan, Japan Society of the Strength and Fracture of Materials, v. 1, p. 309-346.

Rice, J. R., and Cleary, M. P., 1976, Some basic stress diffusion solutions for fluid-saturated elastic porous media with compressible constituents: Review of Geophysics and Space Physics, v. 14 (2), p. 227-241.

Rothman, R. L., 1977, Acoustic emissions in rock stressed to failure: Proceedings of the First Conference on Acoustic Emission/Microseismic Activity in Geological Structures and Materials, Handy, H. R., Jr., and Leighton, F. W., eds., Clausthal, FRG, Trans. Tech., p. 109-133.

Rudnicki, J. W., 1977, The effect of stress-induced anisotropy on a model of brittle rock failure as localization of deformation: Energy Resources and Excavation Technology: Proceedings of the 18th United States Symposium on Rock Mechanics, Colorado School of Mines, Johnson Publishing Company, p. 3B4-1 to 3B4-8.

Rudnicki, J. W., and Rice, J. R., 1975, Conditions for the localization of deformation in pressure-sensitive dilatant materials: Journal of the Mechanics and Physics of Solids, v. 23, p. 371-394.

Rutter, E. H., 1972, Influence of temperature, strain rate and interstitial water in experimental deformation of calcite rocks: Tectonophysics, v. 22 (3-4), p. 311-334.

Salganik, R. L., 1973, Mechanics of bodies with many cracks: Mechanics of Solids, v. 8 (4), p. 135-143.

Scholz, C. H., 1968a, The frequency-magnitude relation of microfracturing in rock and its relation to earthquakes: Bulletin of the Seismology Society of America, v. 58 (1), p. 399-415.

Scholz, C. H., 1968b, Experimental study of the fracturing process in brittle rock: Journal of Geophysical Research, v. 73 (4), p. 1447-1454.

Scholz, C. H., 1968c, Mechanism of creep in brittle rock: Journal of Geophysical Research, v. 73 (11), p. 3295-3302.

Sikarskie, D. L., 1966, Experiments on Wedge Indentation in Rocks: Report TN-262, Princeton, N.J., Ingersol Rand Research and Development Center, 97 pp.

Sikarskie, D. L., and Altiero, N. J., 1973, The formation of chips in the penetration of elastic-brittle materials (Rock): ASME, Journal of Applied Mechanics, ser. E, v. 40 (3), p. 791-798.

Singh, M. M., and Johnson, A. M., 1967, Static and dynamic failure of rock under chisel loads: Transactions AIME, v. 38, p. 366-373.

Sondergeld, C. H., and Estey, L. H., 1981, Acoustic emission study of microfracturing during the uniaxial cycling of Westerly Granite: Journal of Geophysical Research, v. 86 (B4), p. 2915-2924.

Sondergeld, C. H., and Estey, L. H., 1982, Source mechanism and microfracturing during uniaxial cycling of rock: Pure and Applied Geophysics, v. 120 (1), p. 151-166.

Spencer, A. J. M., 1982, Mechanics of Solids: Oxford, Pergamon Press, 607 pp.

Szczepinski, W., 1971, Some slip line solutions for earthmoving processes: Archiwum Mechaniki Stosowanej, v. 23 (6), p. 885.

Tapponier, P., and Brace, W. F., 1976, Development of stress-induced microcracks in Westerly Granite: International Journal of Rock Mechanics, Mineral Science and Geomechanical Abstracts, v. 13 (2), p. 103-112.

Westwood, A. R. C., and Pickens, J. R., 1983, Fracture—applying the breaks: *in* Atomistics of Fracture, Latanision, R. M., and Pickens, J. R., eds., New York, Plenum Press, p. 65-91.

Whittaker, B. N., and Szwilskie, A. B., 1973, Rock cutting by impact action: International Journal of Rock Mechanics and Mineral Science, v. 10 (6), p. 659-671.

Wong, T.-F., 1982, Effects of temperature and pressure on failure and post-failure behavior of Westerly Granite: Mechanics of Materials, v. 1 (1), p. 3-17.

Wong, T.-F., Szeto, H., and Zhang, J., 1992, Effects of loading path and porosity on the failure mode of porous rocks: *in* Micromechanical

Modeling of Quasi-Brittle Materials Behavior, Li, V. C., ed., Book No. AMR 118, New York, ASME, p. 261-398.

Zoback, M. D., Apel, R., Baumgärtner, J., Brudy, M., Emmermann, R., Engeser, B., Fuchs, K., Kessels, W., Rischmüller, H., Rummel, F., and Vernik, L., 1993, Strength of continental crust and the transmission of plate-driving forces: Implications of in-situ stress measurements in KTB scientific borehole, Germany: Nature, v. 365, p. 633-635.

5
ROCK EXCAVATION TOOLS

Introduction

Rock excavation tools disintegrate and remove the rock from boreholes and tunnels by four basic mechanisms: thermal spalling, fusion and vaporization, mechanical stresses, and chemical reactions, as shown in Figure 5.1. "Novel" or "advanced" drilling tools utilize exotic systems such as lasers or electron beams to melt or vaporize rock or explosives, or electrohydraulic discharges, to impact and shatter rock.

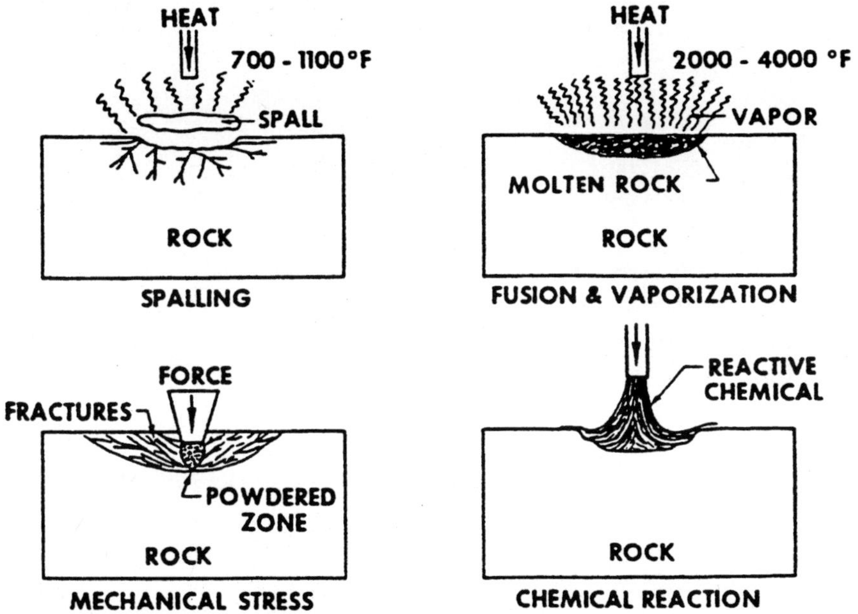

FIGURE 5.1 Basic rock excavation mechanisms (Maurer, 1980).

Status of the Field

Novel Drilling Systems

Thermal spalling drills such as jet-piercing and forced-flame drills heat rocks to 370 to 540°C (700 to 1000°F) to create thermal stresses that spall the rock (Figure 5.2). These devices have limited application because most rocks will not thermally spall (Maurer, 1968, 1980). *Melting and vaporization* drills utilize high-temperature devices such as lasers or electron beams to melt and vaporize rock (Figure 5.3). These devices have relatively low drilling rates because of the high energy requirements to melt and vaporize rock. *Chemical* drills utilize highly reactive chemicals such as fluorine to drill rock (Figure 5.4). These drills have found limited application due to high costs and safety problems associated with handling large volumes of highly reactive chemicals. *Mechanical* stress drills disintegrate the rock by inducing mechanical stresses (Figure 5.5). References for individual novel drills shown in Figures 5.2 to 5.5 (Maurer, 1970) are given at the end of this chapter.

Because the cross-sectional area of a tunnel face is 10 to 100 times greater than a typical drillhole, it is very unlikely that thermal spalling, melting, or vaporization drills could be used as the sole rock removal process due to extremely high power requirements and very low penetration rates. Similarly, chemical drills would not be practical for tunneling because of the large volume of highly reactive chemicals required, safety problems, and problems associated with chemical treatment and disposal of contaminated spoils.

Most of the advanced thermal and high-pressure jet drills require 10 to 100 times more energy to drill rock than conventional rotary bits (Table 5.1). Low drilling rates and excessive power requirements preclude using these advanced devices as the sole rock removal mechanism except in special applications. Additional R&D, which focuses on the physics of the rock removal process, is needed to reduce the overall energy requirements of these novel drilling techniques.

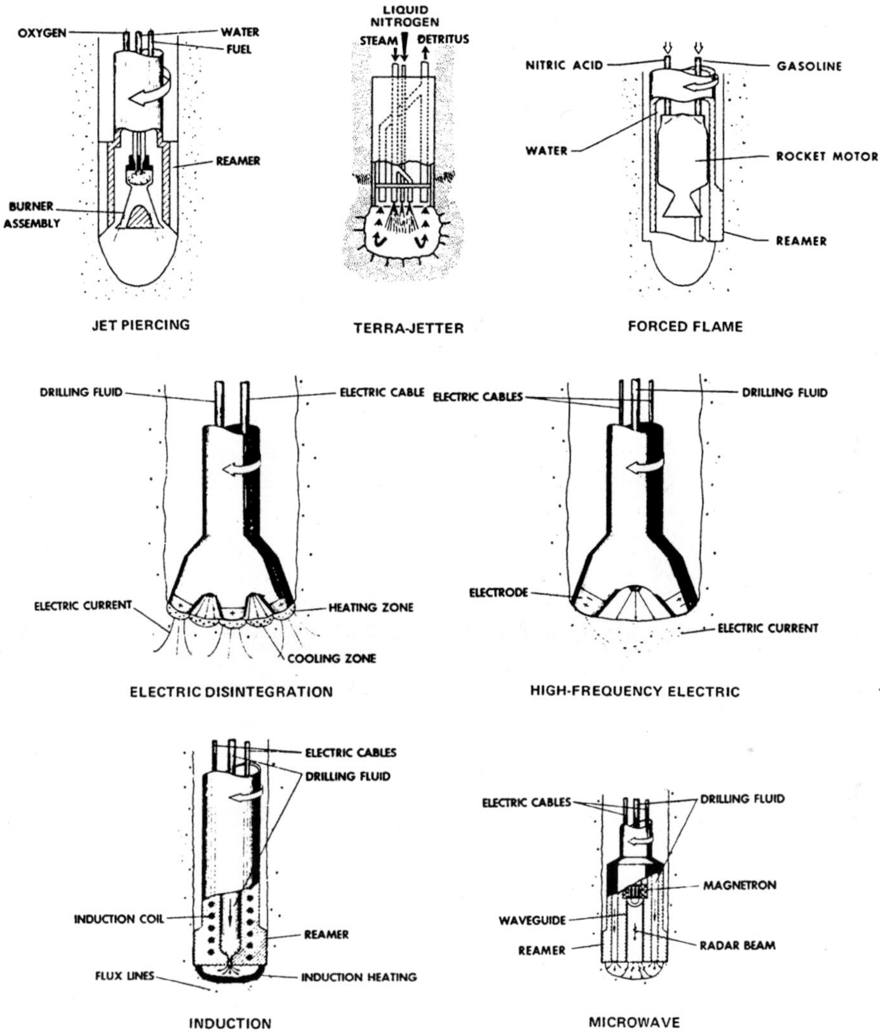

FIGURE 5.2 Thermal spalling drills (Maurer, 1970).

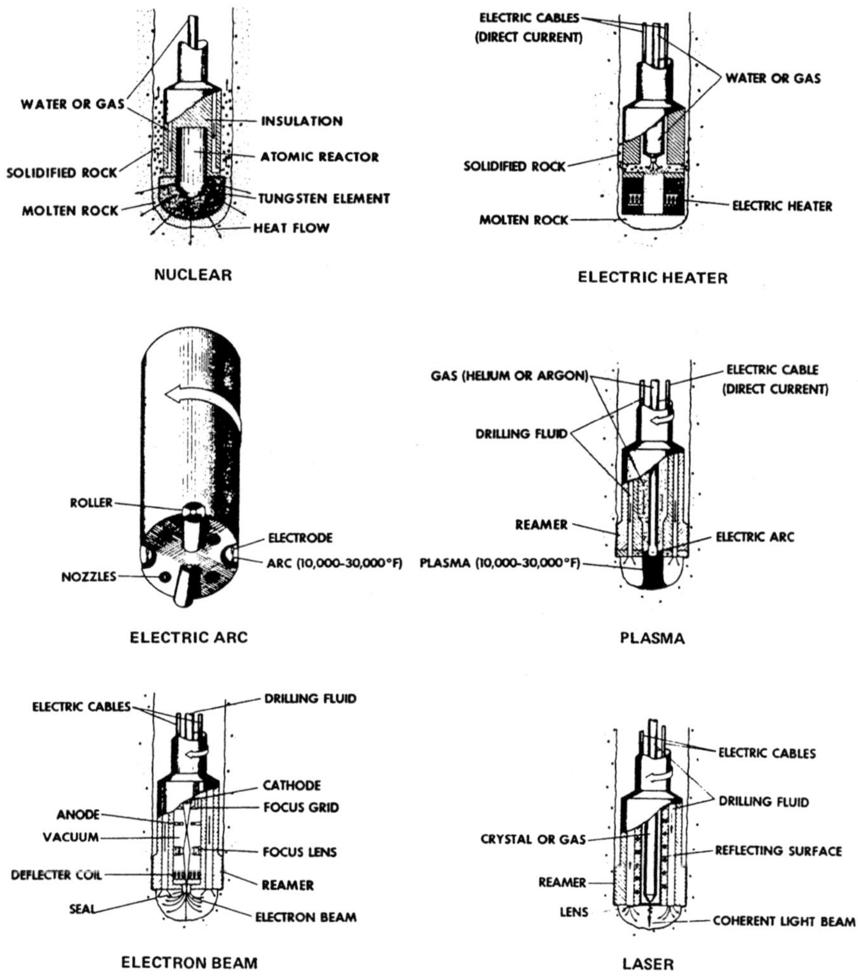

FIGURE 5.3 Melting and vaporization drills (Maurer, 1970).

TABLE 5.1 Specific Energy Requirements for Rock Drilling (Maurer, 1968)

SYSTEM	SPECIFIC ENERGY (joules/cm^3)
Rotary bit	100
High-pressure jets	1,000
Thermal spalling	1,500
Melting	5,000
Vaporization	12,000

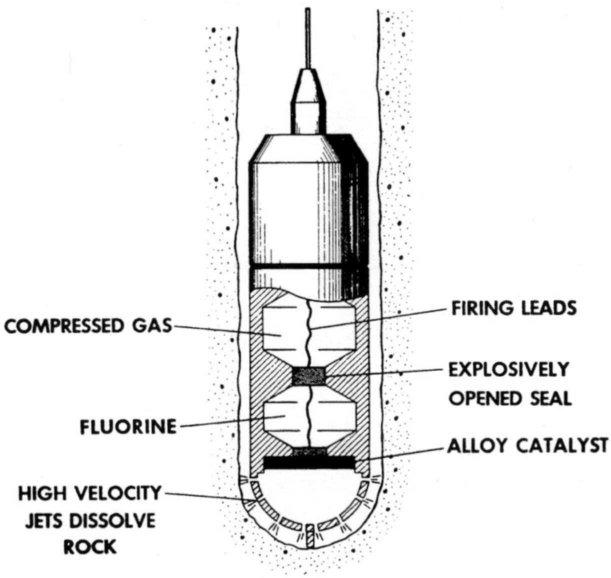

FIGURE 5.4 Chemical drill (Maurer, 1970).

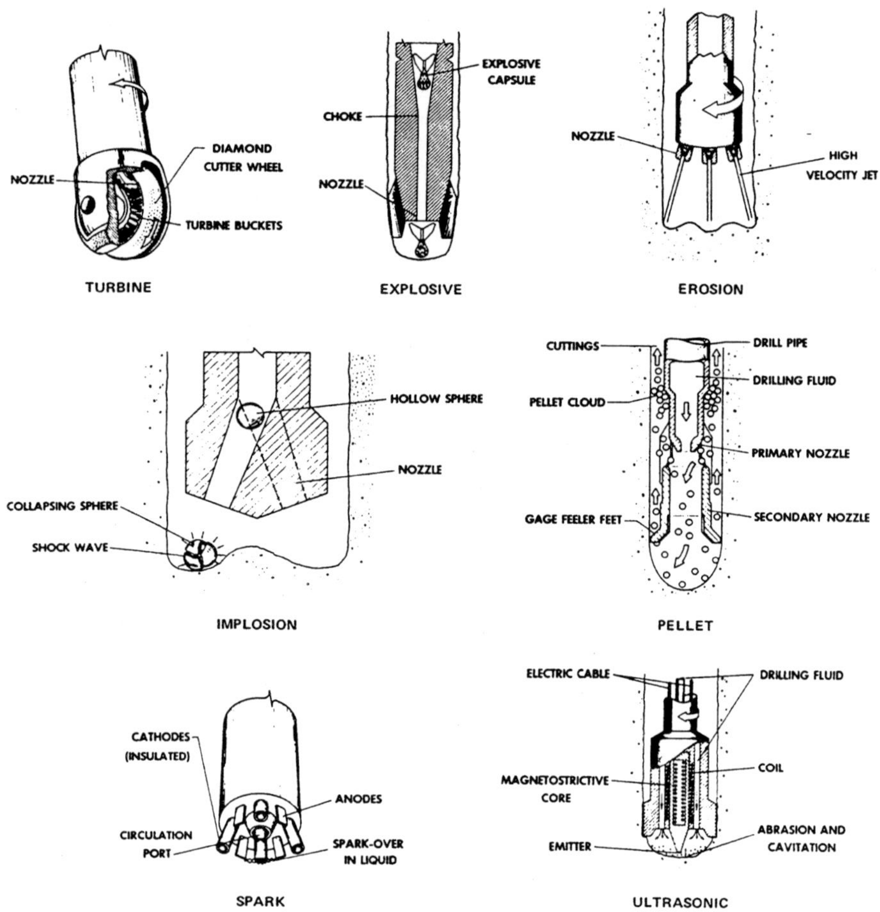

FIGURE 5.5 Mechanical stress drills (Maurer, 1970).

Combined Novel/Mechanical Systems

Because of the high power requirements of novel drilling techniques, advances are more likely to be made on combined novel-mechanical drill bits in which the novel devices (e.g., high-pressure water jets) cut narrow slots or "kerfs" in the rock face, thereby weakening the rock and allowing it to be removed with a conventional bit. These combined systems could yield a two- to fourfold increase in drilling rates (Figure 5.6). A similar system could be used to speed up the advance rate of tunnel boring machines (TBMs; Figure 5.7). Cutting slots in the rock with a novel device weakens the rock and allows conventional cutters to break the rock into larger fragments (Figure 5.8). Figure 5.9 shows an example in which a single slot produced a sevenfold increase in the amount of rock removed by a 75-ft-lb impact. Initial attempts at developing combined novel-mechanical cutters are probably best focused on mining or oil field bits, which are relatively small and inexpensive. Once this technology is developed, it can be scaled up to tunneling and excavation application.

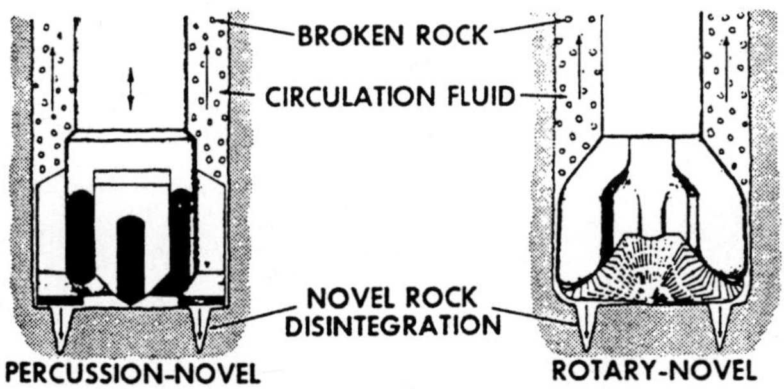

FIGURE 5.6 Combined novel-mechanical drill bits (Maurer, 1980).

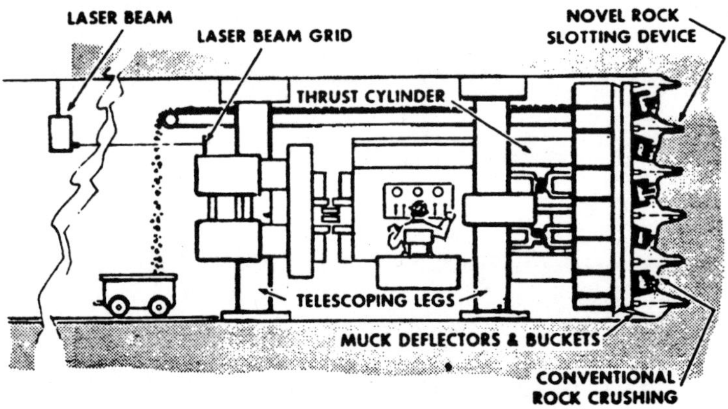

FIGURE 5.7 Novel-mechanical tunnel boring machine (Maurer, 1980).

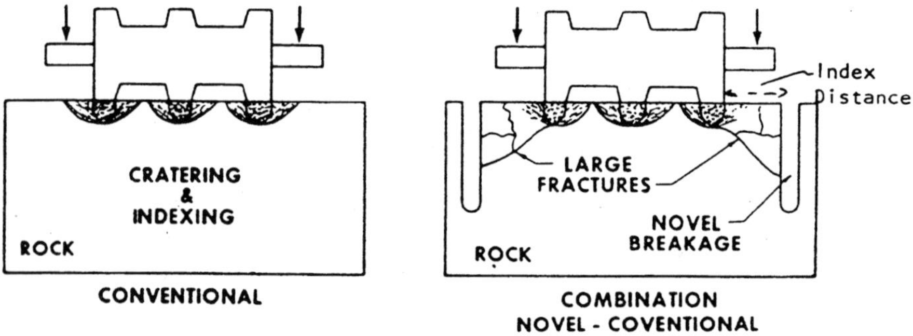

FIGURE 5.8 Combined rock fragmentation (Maurer, 1980).

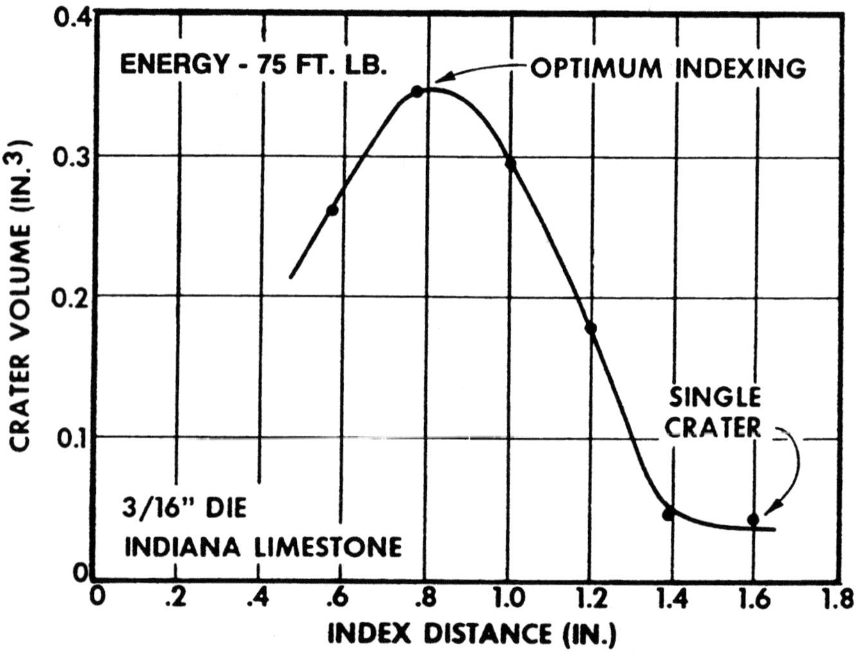

FIGURE 5.9 Effect of index distance on crater volume (Maurer, 1980).

Advanced Downhole Motors

Most guided drilling systems utilize high-speed downhole turbine (Figure 5.10) or positive-displacement motors (200 to 1,000 revolutions per minute [rpm]; Figure 5.11) to obtain high drilling rates. Los Alamos National Laboratory developed geothermal turbodrills that operate effectively in hot-dry-rock geothermal wells at temperatures in excess of 315°C (600°F). These turbodrills accurately guided geothermal holes at drilling rates three to ten times higher than conventional rotary bits. The technology now exists to build downhole motors that will increase drilling rates two- to fourfold by increasing power delivered to the drill bit. Figure 5.12 shows an example (Cohen and others, 1994b) in which a slim-hole motor was overpowered by increasing the fluid flow rate and the differential pressure across the motor. The overpowered motor delivered 56 horsepower (hp) compared to 23 hp for normal operation. In laboratory tests, the overpowered motor drilled marble blocks at rates up to 550 ft/h, compared to 225 ft/h with normal motor operation. Special high-power bits that utilized oversize man-made thermally stable polycrystalline diamond (TSP) cutters were used on this overpowered motor. This example shows that there is significant potential for increasing drilling rates by developing improved motors. Additional R&D is needed on improved air drilling motors and higher-power motors for hard-rock drilling.

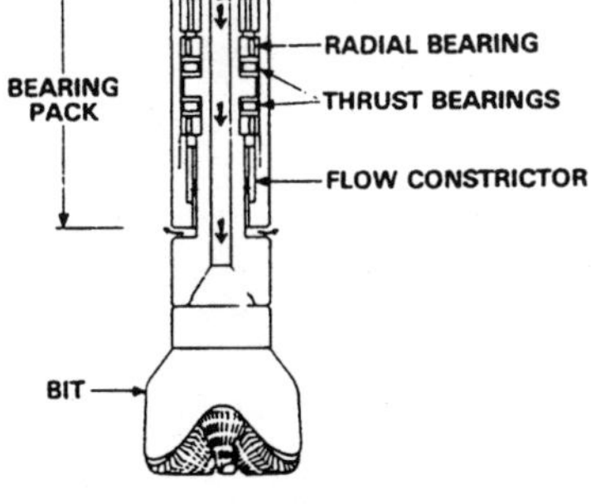

FIGURE 5.10 Downhole turbodrill (Maurer and others, 1978).

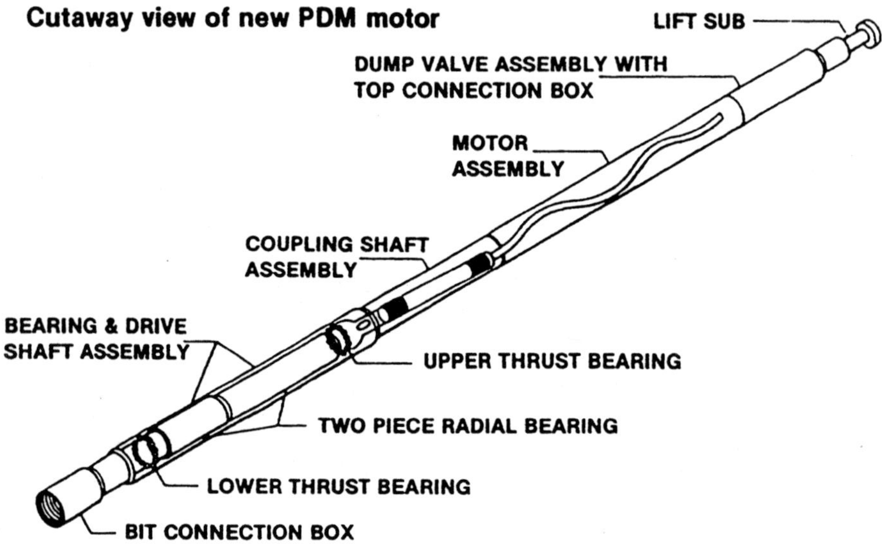

FIGURE 5.11 Cutaway view of new positive displacement mud motor (Dempsey and Leonard, 1979).

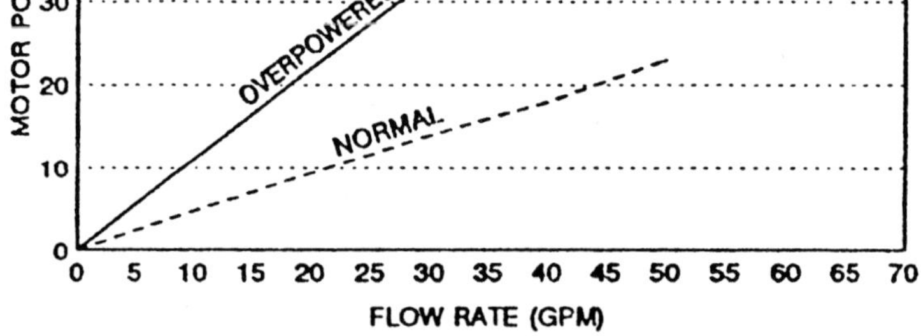

FIGURE 5.12 Overpowered slim-hole motor (Cohen and others, 1994b).

Advanced Drill Bits

Conventional drill bits remove rock by impact or shearing processes (Figure 5.13). Impact or roller bits utilize steel or tungsten carbide cutters to impact and break the rock. New wear-resistant, diamond-coated cutters are finding increased use in hard abrasive rocks. Additional R&D is needed on these improved cutter materials and on improved high-speed bearings.

Shear-type bits utilize polycrystalline or natural diamond cutters to remove rock by shearing processes. Polycrystalline diamond cutter (PDC) bits utilize cutters consisting of a thin layer of small synthetic diamonds bonded to a tungsten carbide substrate, as shown in Figures 5.14 and 5.15. These bits have potential for very high drilling rates because they can operate at much higher rotary speeds (800 to 1,000 rpm) and power levels than roller bits (100 to 200 rpm). In laboratory tests, these bits drill 9 to 14 times faster than conventional roller bits. Further work is needed to develop wear-resistant cutter materials and multipurpose bits that will effectively drill alternating layers of soft and hard rock.

R&D is also needed on sensing tools to look ahead of the bit or a TBM so that potential problems (e.g., blowouts, lost circulation, or waterfilled fractures) can be detected early and appropriate remedial action can be taken. This information can also be used to optimize drilling variables such as the weight on the bit and the rotary speed.

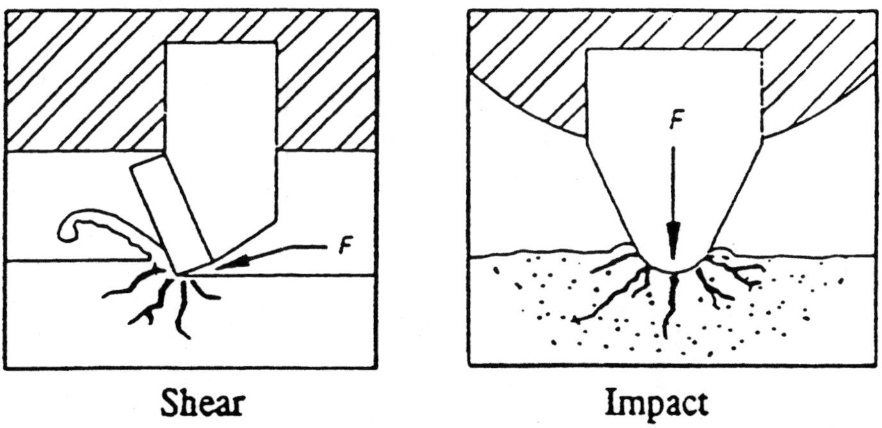

FIGURE 5.13 Drill bit cutting mechanisms.

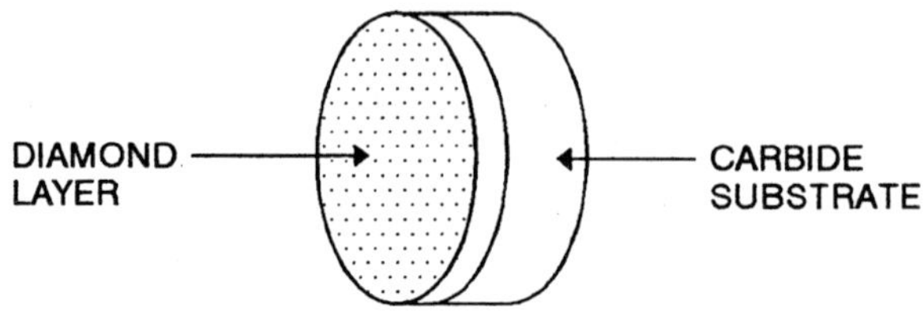

FIGURE 5.14 Polycrystalline diamond cutter (Cohen and others, 1994a).

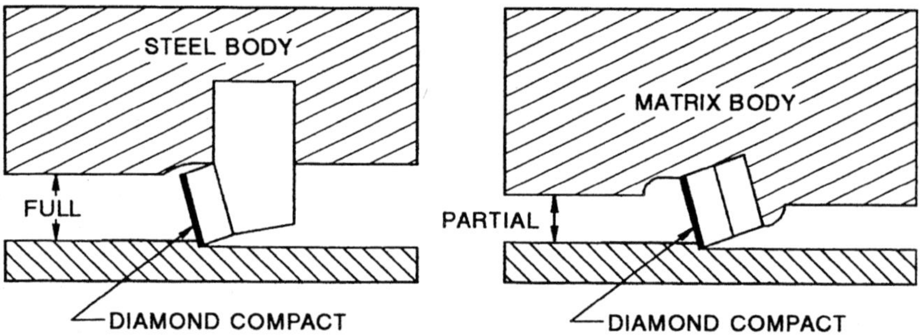

FIGURE 5.15 Polycrystalline diamond cutter mounting techniques (Cohen and others, 1994a).

Advanced TBM Cutters

Tunnel boring machines utilize disk or tooth roller cutters to disintegrate rock (Figure 5.16). The development of the disk cutter by Robbins in the 1980s was a major breakthrough because it allowed much higher cutter-rock contact loads and much higher TBM advance rates in hard rock. R&D on advanced TBM cutters is needed to develop improved cutter materials and longer-life bearings.

Because of its size, a TBM or mining machine excavates two to four orders of magnitude more material per foot of advance than mining or oil field bits. A TBM, therefore, is actually a system of drills, requiring a multiplicity of components. Reliability is thereby reduced because failure of any component stops the entire system.

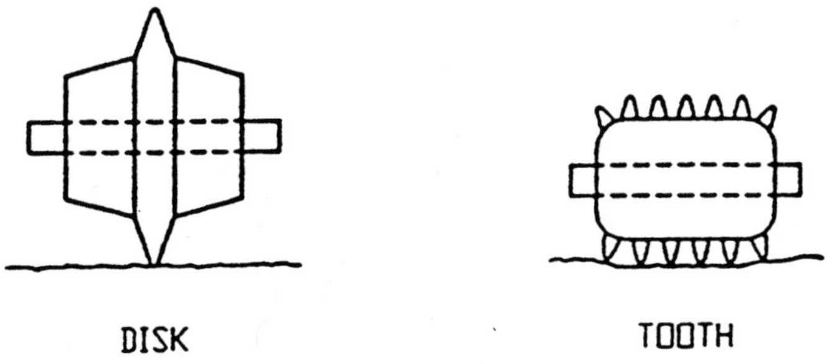

FIGURE 5.16 Tunnel boring machine cutters.

Existing TBMs in soft rock are now capable of such high advance rates that muck removal, rather than rock cutting, is becoming the limiting process for the system. Thus, significant improvements in excavation will require parallel development of both the rock cutting and the muck removal components of the system.

High-Pressure Jet Drilling Systems

Experimental high-pressure jet drilling systems operating at 10,000 to 20,000 pounds per square inch (psi) can drill two to three times faster than conventional drills (Figures 5.17 and 5.18). A dual-wall drill-pipe system currently being field tested utilizes mud pumped at high pressure (34,000 psi) to cut slots in the rock ahead of roller bits and lower-pressure mud to remove the broken rock from the hole bottom (Figure 5.19). This system drilled two to three times faster than conventional bits in an east Texas oil field test (Figure 5.20). In another experiment, a high-pressure downhole motor operating at 10,000 psi produced drilling rates in excess of 1,000 ft/h in medium-strength rock.

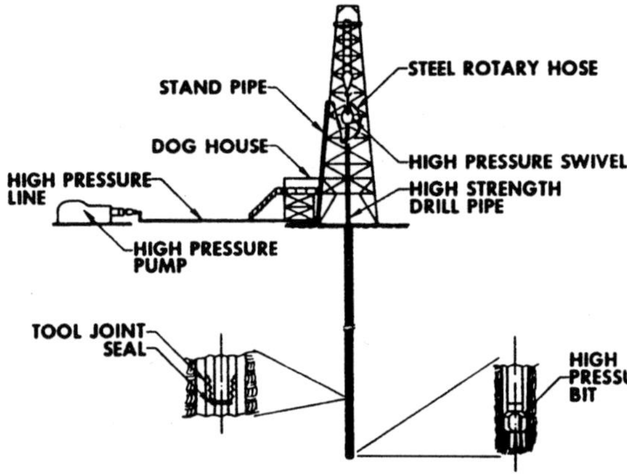

FIGURE 5.17 Exxon high-pressure rig (Maurer and others, 1973).

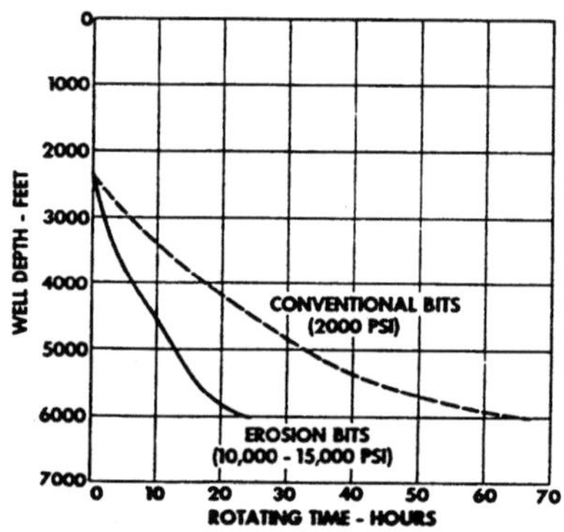

FIGURE 5.18 Exxon drilling data for high-pressure versus conventional bits (Maurer and others, 1973).

Despite these high drilling rates, these systems have not been commercialized because of problems with the high-pressure equipment. These problems include erosion and leaks in the drill pipe caused by the abrasive drilling muds. A new concept being evaluated to eliminate these problems is the use of high-pressure downhole pumps powered by low-pressure pumps at the surface (Figure 5.21). Additional work on these systems is needed to develop improved pumps, drilling pipe, high-pressure motors, and jet bits.

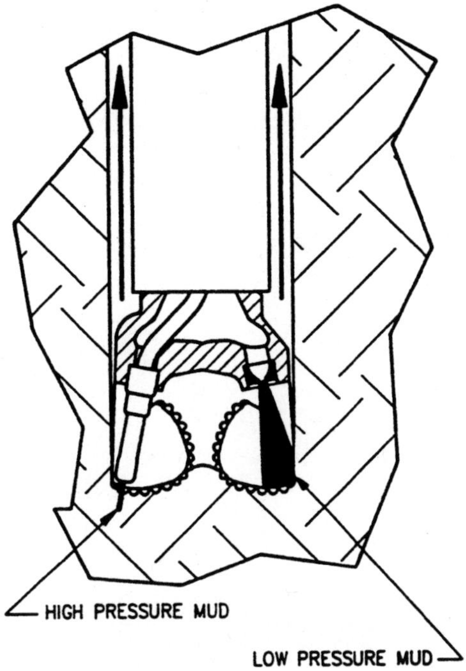

FIGURE 5.19 FlowDril jet bit (Butler and others, 1990).

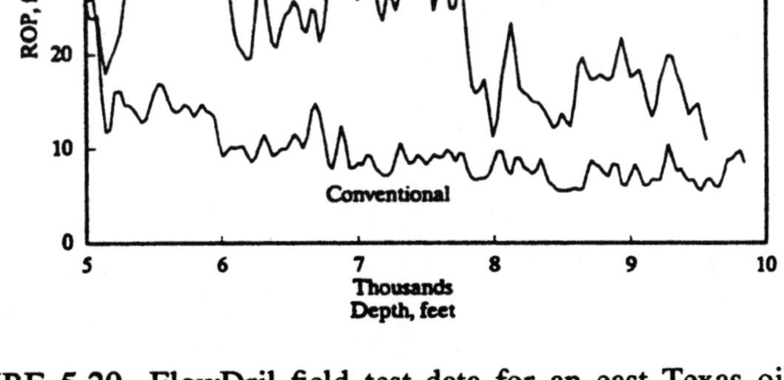

FIGURE 5.20 FlowDril field test data for an east Texas oil field test (Kolle and others, 1991).

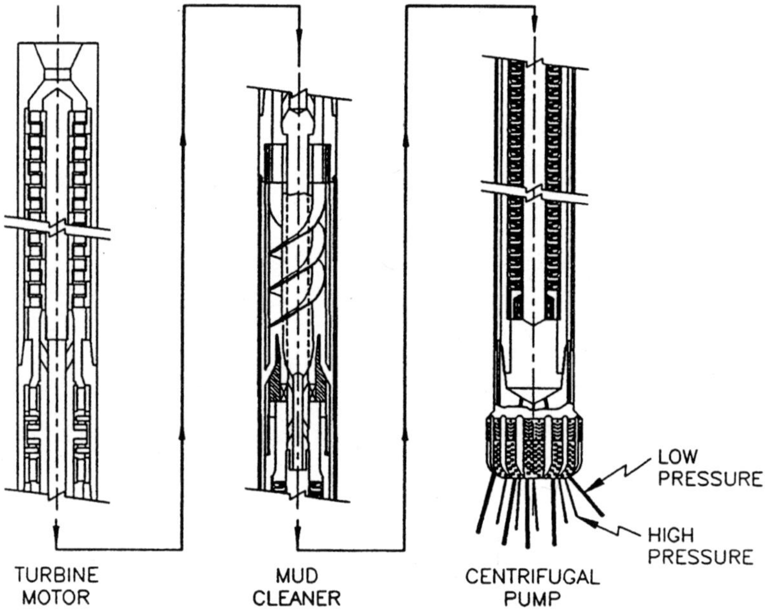

FIGURE 5.21 High-pressure downhole pump.

Slim-Hole Drilling Tools

Slim-hole drilling is finding increased use in the oil and gas industry, because it lowers well costs by 40 to 70% due to reduced materials (mud, cement, and casing), smaller rigs and crews, and faster drilling rates. Improved slim-hole drilling tools are needed because space constraints limit the strength and life of these tools.

Guided Percussion Drills

Percussion or hammer drills can penetrate many hard rocks two to four times faster than rotary drills because they apply high-impact loads that shatter the rock into large fragments. Although widely used in shallow blast-hole operations, percussion drills have found limited use for deep oil, gas, and geothermal drilling because of the inability to accurately guide these drills in hard rock. Improved percussion drills that can be guided accurately in hard rock are needed for deeper drilling applications.

Resonant Sonic Drills

Resonant sonic drills (Volk and others, 1993) use eccentric rotating weights to vibrate the drill pipe at frequencies of 80 to 150 cycles per second to fluidize soil and other unconsolidated materials, thus allowing the pipe to be advanced into the ground with minimal friction. No fluid circulation is used with these drills, so the material ahead of the bit either is pushed aside into the surrounding formation or flows into the core barrel and is retrieved back to the surface.

Sonic drills have been around since the 1950s but have never caught on owing to their relatively high cost and poor reliability (the tool is particularly susceptible to vibrational fatigue). Efforts are now under way at Hanford and Sandia Labs (Volk, 1992; Volk and others, 1993) to improve the reliability of this tool for use in environmental drilling and coring. Sonic drills are currently limited to drilling primarily unconsolidated materials to maximum depths of about 400 ft. Methods are needed to guide these drills directionally.

Priorities for R&D

As noted in Chapter 4, the fundamental rate-controlling process in drilling is the rate of rock breaking by the drilling tool. Significant improvements in rock breaking rates can be achieved through a better understanding of the physics of rock-tool interactions. Revolutionary advances in rates of rock breaking and rock removal are possible through the development of hybrid mechanical-novel drilling tools to break and remove rock. Evolutionary advancements in rock removal rates are likely through the development of stronger, wear-resistant cutters, bearings, and bits. R&D should be focused in the following areas:

1. *Physics of rock-tool interactions:* The physical and mechanical characteristics of rocks are generally well known, but much less is known about the mechanics of rock removal, that is, the interaction between the rock and the drilling tool. Research that focuses on the physics of rock

removal processes is a prerequisite for designing improved drilling tools and should have a high priority for support.

2. *Improved cutter materials and bearings:* Conventional drill bits utilize steel or carbide cutters to remove rock by shearing or impact. Wear-resistant, diamond-coated cutters are finding increased application in hard, abrasive rocks, particularly for shear-type bits. Additional R&D is needed on these wear-resistant cutter materials, and on these and other wear-resistant materials for high-speed bearings.

3. *Novel-hybrid drilling technologies:* R&D on novel drilling technologies should focus on reducing energy requirements for drilling. The development of hybrid systems, in which novel drilling tools are used to lower the strength of the rock, and conventional mechanical drilling tools are used to break and remove it, are especially promising and should be pursued. R&D on novel systems should focus initially on mining and oil field bits, which are relatively small and inexpensive. Once this technology has been developed, it should be scaled up to tunneling and excavation applications.

4. *Improved mechanical drills:* Significant improvements in rock removal rates can also be obtained through evolutionary advances in conventional mechanical drilling tools. R&D efforts should focus on the development of the following:

- high-speed, high-power downhole motors;
- guided percussion drills; and
- slim-hole drilling tools.

References

Agoshkin, M. I., and Vornyuk, A. S., 1960, Secondary breaking of iron ores by high-frequency current: Izvestiya Akademii Nauk USSR, no. 1, p. 138-144, English Translation No. 4857 by Henry Brutcher Technical Translations, P.O. Box 157, Altadena, Calif.

Angona, F. A., 1971, Method and Apparatus for Explosive Drilling Utilizing Spark Pumps for Detonating Explosives: U.S. Patent No. 3,576,219, April 27, 6 pp.

Anon., 1961, New role for R-F energy: rock bursting: Electronics, September 22, p. 11.

Anon., 1963, Drilling with radar: Mining and Engineering Journal, April, p. 24.

Anon., 1973, New soviet rocket sinks wells, Industrial Research, October, p. 35.

Avery, R. T., Keefe, D., Brekke, T. L., and Finnie, I., 1973, Shattering rock with intense bursts of energetic electrons: IEEE Transactions on Nuclear Science, v. NS-20 (3), p. 1010-1014.

Baum, W., 1978, Burners for Flame Jet Drill: U.S. Patent 4,073,351, February 14, 8 pp.

Benson, G. M., 1977, Thermocorer for rapid tunneling and excavation: Proceedings of the 1977 NSF Research in Excavation Technology Conference, Houston, Texas, April 26-28, Oakland, Calif., Energy Research and Generation, 125 pp.

Bergstrom, B. H., 1961, The electrohydraulic crusher: Engineering and Mining Journal, v. 162 (2), p. 134-140.

Black, W. L., 1974, Basic Understanding of Earth Tunneling by Melting: Report No. DOT-TSC-OST-74-6.1, prepared for the U.S. Department of Transportation by Westinghouse Astronuclear Laboratory, Pittsburgh, Pa. (2 volumes).

Black, W. L., 1976, Development of Equipment for Explosive Drilling: AAI Corporation, Cockeysville, Md., Engineering Report ER-8724, DOD Contract DAAG56-76-C-0033, 37 pp.

Bobo, R. A., 1969, Method of Drilling with High Velocity Jet Cutter Roller Bit: U.S. Patent 3,112,800, December 3, 7 pp.

Brichkin, A. V., and Bolotov, A. V., 1966, Use of an electric arc for rock boring: Izvestiia Vysshikh Uchebnykh Zavedenii, Tsvetsnaya Metallurgiya USSR, May.

Browning, J. A., 1961, Method and Apparatus for Impacting a Stream at High Velocity Against a Surface to be Treated: U.S. Patent No. 2,990,653, 9 pp.

Browning, J. A., and Fitzgerald, E. M., 1969, Method of Flame Drilling with Abrasives: U.S. Patent No. 3,463,249, August 26, 5 pp.

Butler, T., Fontana, P., and Otta, R., 1990, Method for combined jet and mechanical drilling: SPE 20460, Drilling Proceedings, SPE Annual

Technical Conference and Exhibition, Richardson, Tex., Society of Petroleum Engineers, v. delta, p. 561-565.

Calaman, J. J., and Rolseth, H. C., 1962, Technical advances expand use of jet-piercing process in taconite industry: Mining Research, Oxford, Pergamon, p. 473-498.

Carstens, J. P., Davison, W. R., Brown, C. A., McGarry, F. J., and Smith, A. R., 1970, Heat-Assisted Tunnel Boring Machines: United Aircraft Research Laboratories, East Hartford, Conn., September (NTIS-PB-197, 243), 335 pp.

Carstens, J. P., and Brown, C. A., 1971, Rock Cutting by Laser: SPE 3529, 46th Annual Meeting of the Society of Petroleum Engineers, New Orleans, La., October 3-6, Richardson, Tex., SPE.

Clark, I. E., and Shafto, G. R., 1987, Core drilling with SYNDAX3 PCD: Industrial Diamond Review, v. 47 (251), p. 169-173.

Cohen, J. H., Maurer, W. C., and Westcott, P. A., 1994a, High-power TSP bits: Journal of Energy Resources Technology, v. 116 (Mar.), p. 33-37.

Cohen, J. H., Maurer, W. C., and Westcott, P. A., 1994b, Improved high-power TSP bits: *in* Drilling Technology, Vonzia, J. P., ed., presented at Energy-Sources Technology Conference & Exhibition, New Orleans, La., January 23-26, v. 56, p. 139-144.

Conn, A. F., and Radtke, R. P., 1977, Cavitating bit jets promise faster drilling for deep-hole operations: Oil and Gas Journal, v. 75 (Oct. 31), p. 129-146.

Cooley, W. C., 1972, Rock breakage by pulsed high pressure liquid jets: 1st International Symposium on Jet Cutting Technolgy, BHRA Fluid Engineering, Cranfield, Bedford, England.

Dardick, D., 1977, Ground Terra-Drill Processes and Apparatus: U.S. Patent No. 4,004,642, January 25, 9 pp.

Dempsey, P., and Leonard, J., 1979, Mud motor is tested: Drilling Contractor, May, p. 54-57.

Dodd, H. M., 1976, An Economic Analysis of Downhole Replaceable Drill Bit Systems: Sandia Laboratories Report No. SAND76-0301, Albuquerque, N. Mex., 55 pp.

Epperson, J. P., Dyer, W., and Grzywa, J. C., 1966, The laser now a production tool: Proceedings of the 8th Annual Electron and Laser Beam Symposium, Haddad, G. I., ed., Ann Arbor, University of Michigan, p. 187-205.

Epshteyn, E. F., Arsh, E. I., and Vitort, G. K., 1960, New methods of crushing rock: Gosud, Naushno-Tekh, Izad, Neft. i Gorno-Topliv Noi Literatery Moscow, Translation No. 62-11712, Washington, D.C., U.S. Department of Commerce, Office of Technical Services, 85 pp.

Frankle, H., and Baum, W., 1978, Flame Jet Tool for Drilling to Great Depths: U.S. Patent No. 4,099,584, July 11, 9 pp.

Hartman, H. L., 1963, The simulation of percussion drilling in the laboratory by indexed-blow studies: SPE 500, Proceedings of the First Conference on Drilling and Rock Mechanics, University of Texas, Austin, Tex., January 23-24, Richardson, Tex., Society of Petroleum Engineers.

Jurewicz, B. R., Carstens, J. P., and Banas, C. M., 1974, Rock-kerfing with high-power lasers: Proceedings of the 3rd Congress of the International Society of Rock Mechanics, Denver, Colo., September 1-7, National Academy of Sciences, v. 12, pt. B, p. 1434-1440.

Jurewicz, B. R., Greenwald, L. E., and Brown, C. O., 1974, Focused Laser Beams to Assist Rock Excavation: United Aircraft Research Laboratories, East Hartford, Conn., DOT Contract No. FR-20021, 166 pp.

Kolle, J. J., Otta, R., and Stang, D. L., 1991, Laboratory and Field Testing of an Ultra-High Pressure, Jet-Assisted Drilling System: SPE/IADC 22000, 1991 SPE/IADC Drilling Conference held in Amsterdam, March 11-14, Richardson, Tex., SPE, p. 847-856.

Kravchenko, V. S., 1957, The search for new methods of breaking down hard rock: Gornyi Zhurnal USSR, no. 1, p. 36-43.

Maroudas, N. E., Johnson, H. A., and Yigit, E., 1967, The mechanism of electrohydraulic crushing: Dechema-Monographien, v. 57, p. 551-582.

Marovelli, R. L., Chen, T. S., and Veith, K. F., 1966, Thermal fragmentation of rock: Society of Mining Engineers, Transactions, v. 235 (1), p. 1-15.

Maurer, W. C., 1968, Novel Drilling Techniques: Oxford, Pergamon Press, 114 pp.

Maurer, W. C., 1969, Spark drilling: Proceedings of the 11th Symposium On Rock Mechanics, University of California, Berkeley, Calif., June 16-19, Houston, Tex., Society of Mining Engineers, AIME, p. 687-703.

Maurer, W. C., 1970, Potential application of novel rock disintegration techniques in mining: presented at the 1970 Joint Minnesota Section,

AIME Annual Meeting and University of Minnesota Mining Conference, Duluth, Minn., January 12-14, New York, AIME, 45 pp.

Maurer, W. C., 1980, Advanced Drilling Techniques: Tulsa, Oklahoma, Petroleum Publishing, 698 pp.

Maurer, W. C., and Heilhecker, J. K., 1969, Hydraulic jet drilling: SPE 2434, Proceedings of 4th Conference on Drilling and Rock Mechanics, University of Texas, Austin, Tex., January 14-16, Richardson, Tex., SPE.

Maurer, W. C., Heilhecker, J. K., and Love, W. W., 1973, High-pressure drilling: Journal of Petroleum Technology, v. 25 (7), p. 851-859.

Maurer, W. C., Rowley, J. C., and Clifton, C., 1978, Advanced turbodrills for geothermal wells: Geothermal Resources Council, Transactions, v. 2, p. 411-414.

Neppiras, E. A., 1959, Ultrasonic machining: Modern Workshop Technology, Part 2, London, Cleaver-Hume Press, p. 217-228.

Newsom, M. M., 1975, Status Report—Drilling Research on the Electrical Detonation and Subsequent Cavitation in a Liquid Technique (Spark Drilling): Sandia Laboratories Report No. SAND-75-0417, Albuquerque, N. Mex., 59 pp.

Ostrovskii, N. P., 1960, Deep-hole drilling with explosives: Gostoptekhia 'dat Moscow, Translation by Consultants Bureau Enterprises, Inc., New York.

Peterson, C. R., 1973, Study of a Continuous Drill-and-Blast Tunneling Concept: Rapidex, Inc., Final Report of an Advanced Research Project Agency-Supported Study, Boxford, Mass., RAPIDEX, 55 leaves.

Pols, A. C., 1977, Tests show jet-drilling hard-rock potential: Oil and Gas Journal, v. 75 (Jan. 31), p. 134-144.

Robinson, L. H., Jr., 1962, Drilling Boreholes with Explosive Charges: U.S. Patent No. 3,070,010, December 25, 9 pp.

Robinson, L. H., 1965, Experimental tests of a method for drilling with explosives: Society of Petroleum Engineers Journal, June, p. 153-159.

Rowley, J. C., 1974, Rock melting applied to excavation and tunneling: Proceedings of the 3rd ISRM Congress, Denver, Colo., September 1-7, National Academy of Sciences, Washington, D.C., National Academy Press, p. 1447-1453.

Sarapuu, E., 1963, Electrical disintegration drilling: *in* Rock Mechanics, Fairhurst, C., ed., Oxford, Pergamon, p. 173-184.

Sarapuu, E., 1969, Breaking rocks with electric current: The Australian Miner, June 30.

Schumacher, B. W., 1969, Electron beams apply an old principle to modern rock breaking: Engineering and Mining Journal, June, p. 116-120.

Schumacher, B. W., and Holdbrook, R. G., 1972, Use of Electron Beam Gun for Hard Rock Excavation—Final Report: ARPA Contract No. H0110377, December 30 (NTIS-AD-761208), Sunnyvale, Calif., Westinghouse Electric.

Singh, M. M., and Gregson, V. G., Jr., 1969, Rock Breakage by High Gas Gun Projectiles: Office of High Speed Ground Transportation, Deptartment of Transportation Report (Under Contract DOT 3-0170), Chicago, IIT Research Institute, 129 pp.

Stewart, A., Falter, F. X., and Tomlinson, P. N., 1988, Drilling reef quartzite with SYNDAX3: Industrial Diamond Review, v. 48 (526), p. 117-119.

Summers, D. A., and Henry, R. L., 1972, Water jet cutting of sedimentary rock: Journal of Petroleum Technology, v. 24 (7), p. 797-802.

Thirumalai, K., 1970, Rock Fragmentation by Creating a Thermal Inclusion with Dielectric Heating: Bureau of Mines RI 7424, 33 pp.

Thorpe, M. L., 1960, The plasma jet and its uses: Research/Development, v. 11 (1), 15 pp.

Titkov, N. I., and others, 1957, Drilling with electrical discharges in liquids: Neftyanoe Khozyaistvo USSR, English Translation by Associated Technical Services Inc., P.O. Box 271, East Orange, N.J.

Volk, B. W., 1992, Results of Testing the Sonic Drilling System at the Hanford Site (September 1991 to May 1992): Report WHC-SD-EN-TRP-002, prepared for the U.S. Department of Energy by Westinghouse Hanford Company, Richland, Wash.

Volk, B. W., McLellan, G. W., Moak, D. J., Lerch, R. E., Thompson, K. M., and Barrow, J. C., 1993, ResonantSonic Drilling: History, Progress and Advances in Environmental Restoration Programs: Report WHC-SA-1959-FP, prepared for the U.S. Department of Energy by Westinghouse Hanford Company, Richland, Wash., 21 pp.

Voytsekhovskiy, B. V., 1964, Some results of the destruction of rocks by means of a pulsed water jet: Izvestiya Sibirskogo Otdeleniya Akademii Nauk USSR, English Translation No. OTS: 64-21809, U.S. Department of Commerce, Office of Technical Services, Washington, D.C.

Voytsekhovskiy, B. V., 1967, Jet Nozzle for Obtaining High Pulse Dynamic Pressure Heads: U.S. Patent 3,343,794, September 26, 4 pp.

Watson, J. D., 1974, Applications of Military Guns to High Speed Drilling and Excavation in Rock—Project REAM: Physics International Report No. PIFR-551, prepared for the USBM, Physics International, 45 leaves.

Wyllie, M. R. J., 1971, Jetted particle drilling: Proceedings of Eighth World Petroleum Congress, Moscow, USSR.

Young, F. M., 1963, The secondary breaking effect of heat-frequency electric energy applied to rock fragments: *in* Rock Mechanics, Fairhurst, C., ed., Oxford, Pergamon Press, p. 185-203.

Yutkin, L. A., 1955, Electrohydraulic Effect: Moscow, Mashgiz (State Scientific Technical Press for Machine Construction Literature), Translation No. AD-267-722, Armed Services Technical Information Agency, Arlington Hall Station, Arlington, Va.

6
SENSING SYSTEMS

Introduction

To support the development of smart drilling systems, incremental advances that can come from a systems approach to the entire rock breaking and removal system must be recognized (Onoe and others, 1991). Yet, the intelligence behind the smart drilling system lies in the sensing systems to be incorporated in each drilling system component. Sensing systems include devices that measure and analyze data, interpret results, and activate other systems in response to the interpreted results. Functions of sensing systems include measurements of the drill or tunnel borer, measurements of the geologic formation, and measurements of the interaction between drill bit and rock, as well as positioning and telemetry. Sensing systems analyze measurements by human intervention, as well as by completely autonomous expert-driven systems.

Sensing on drilling systems has rapidly evolved over the last 20 years. Advances in the oil and gas drilling area, discussed here, generally pertain to other drilling technologies as well. Simple sensing of the rate of rotation, the rate of penetration, the torque, and the weight on the drill string, all measured at the surface, has gradually been enhanced to include several measurements of formation properties, in addition to downhole analysis of drilling parameters. These measurements are commonly made 30 to 50 ft behind the drill bit in special sections of the drill pipe. The data are analyzed by downhole computers, and because the drill pipe sections do not form a usable continuous electrical circuit, the final results are transmitted to the surface by pressure pulses in the mud column.

The array of formation measurements that is available while drilling is now approaching the set of standard wireline data that are collected after the drill string is removed, and this trend will probably continue (Cantrell and others, 1992). Therefore, it is useful to consider wireline data as an analogue of what may be done in the future by a smart drilling system.

Different types of measurements may be made by using sondes (i.e., instrument packages) suspended on a wireline cable after the drill string has been removed (Figure 6.1). The physics of these instruments is reviewed by Tittman (1986) and Ellis (1987), and an insight into geological interpretation is given by Doveton (1986). The measurements are conveniently grouped into electromagnetic, nuclear, acoustic, and mechanical categories. Usually the instrumentation consists of a source device, a receiver, and electronics to digitize and transmit the data to the surface. For measurements of some properties, such as the Earth's naturally occurring radioactivity, no source is required.

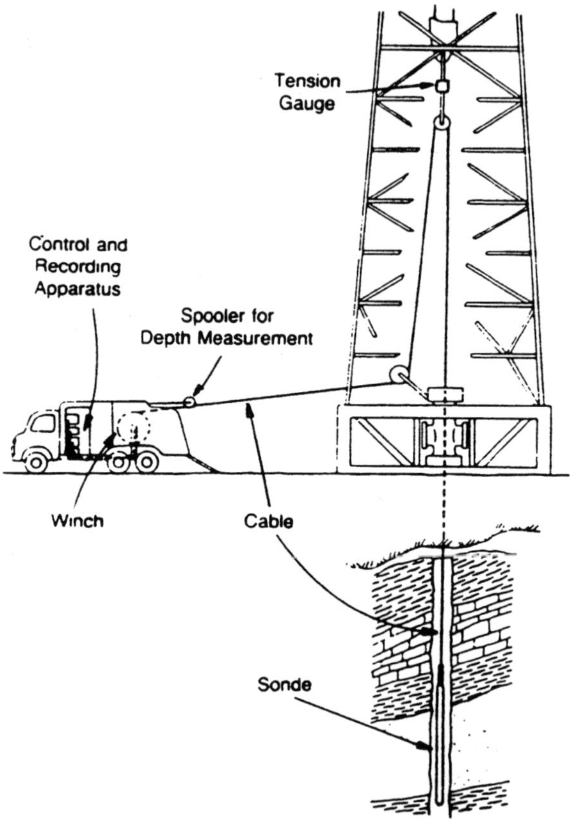

FIGURE 6.1 Generalized logging setup (Desbrandes, 1982).

Electromagnetic devices are used to distinguish resistive oil and gas from conductive brines in rock pore spaces. An example of an electromagnetic device is shown in Figure 6.2. Most rock is also resistive, however, and the overall resistivity depends on the porosity, that is, the volume of pore space as well as the conductivity of the enclosed fluids. The spontaneous potential log is typically used to estimate brine conductivity. Porosity can be estimated concurrently with conductivity by using nuclear and acoustic methods. The velocity of sound depends largely on the porosity and secondarily on the rock type. The scattering of neutrons or high-energy gamma rays from a source to a detector depends primarily on the porosity and secondarily on the rock type, and sondes have been constructed based on both properties.

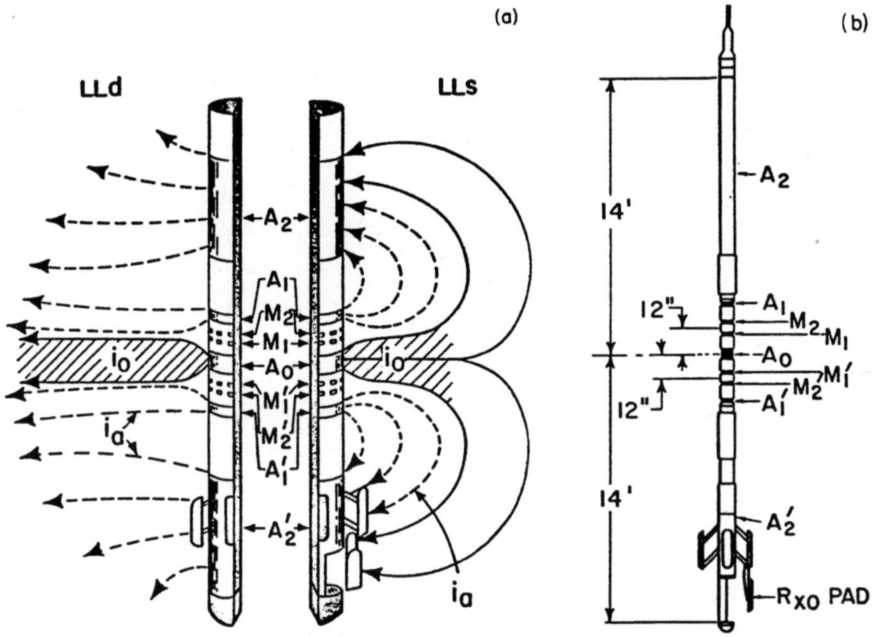

FIGURE 6.2 (a) Electrodes and current distribution for the Dual Laterolog in an infinite and homogeneous isotropic medium. The sonde is split for purposes of illustration. (b) Arrangement of electrodes on the sonde (Suau and others, 1972).

Rock type is determined from the combination of porosity-sensitive measurements and chemical composition from natural or induced gamma rays. Additional, more specialized measurements are available to aid in the determination of sedimentary structures, fractures, faults, and pore-size distributions. Ultimately, the information available from wireline sondes enables a comprehensive characterization of the rock as well as the type, amount, and ultimate recovery of hydrocarbons.

Although not as extensive as the available suite of wireline measurements, the set of measurements made while drilling is impressive (Gianzero and others, 1985; Jan and Harrel, 1987; Norve and Saether, 1989; Wraight and others, 1989). These measurements include parameters related to lithology, porosity, and hydrocarbon identification. Downhole formation measurements are being combined with one another, creating new opportunities in drilling efficiency (Betts and others, 1990). Measured formation properties can now be compared continuously to values expected for the target horizon, as long as the drill remains in the target formation, the sets of parameters agree, and the drilling trajectory is unaltered. However, when measured parameters deviate from expected values, it is likely that the drill is no longer in the target formation. The drilling trajectory can then be modified to remain within the target or to reenter the target zone. Recent advances in these technologies show that even more efficient drilling practices could be achieved if the intelligence of the drilling system is increased through the use of advanced sensing systems.

Status of the Field

In what follows the major components and principal problem areas of sensing systems are considered. The major components are bit-rock interaction; formation evaluation and fracture detection at and ahead of the bit; measurement of stress, rock strength, and pore pressure; drill pipe-rock interaction; position, direction, and steering analysis; cutting and mud analysis; environmental logging; telemetry; and data analysis, measurement interpretation, and activation systems.

Bit-Rock Interaction

Current measurements at the critical bit-rock interface are limited to the rate of penetration, the weight on the bit, the torque, and the rate of rotation. This set of measurements is also used to estimate bit wear (Fay, 1993). It is possible that these bit-rock interface measurements can be significantly enhanced through new discoveries in the fundamental physics and chemistry of bit-rock interaction. Advances now allow measurements at the bit that can establish the particular rock breaking mechanism that is active at that moment (Roy and Cooper, 1993). Other advances now allow monitoring of the state of wear of the bit from bit-emplaced sensors. Information gained from such monitoring indicates that the bit being used is not the most efficient under current conditions and that bit replacement, change in operating conditions, or other remedial action is called for. One recent development in this area is the Ocean Drilling Program's development of a cutter head with retractable bits that can be changed without removing the cutting head.

Formation Evaluation and Fracture Detection at and Ahead of the Bit

The measurement of physical and chemical (e.g., contaminants) properties at the drill bit and the prediction of rock properties ahead of the bit hold enormous potential. A great deal of progress has been made over the past decade in measurement-while-drilling (MWD) technology. Before MWD, one had to wait until the drill string was removed to log or directly observe formations for their lithology and mineral or hydrocarbon content. The time lag sometimes lasted several days. With MWD, sensing systems are placed in the drill string above the bit. Information about the formations encountered uphole is recorded and transmitted with the drill string in place hours after drilling.

Current MWD systems enable sophisticated formation evaluation a few meters behind the current bit location by measuring a wide array of formation properties including resistivity, natural gamma-ray activity, bulk density, and neutron porosity. It is likely that in the near future the set of MWD measurements will expand to include most routine wireline logging data.

Advances in MWD now reduce the distance lag behind the bit where measurements are currently made and expand the scope to include measurements ahead of the bit. Such measurements would hold great promise for many industries and applications. For example, in environmental logging, fractures carry contaminants into or from sensitive areas. In tunneling, open, water-bearing fractures cause enormous problems. Not only are the tunnelers immediately endangered, but the remedial cost is significant. It would be extremely beneficial to be able to detect and plan for such conditions before they are encountered by the bit. Another example involves oil and gas drilling where high-pressure gas kicks can "blow out" a well. If such high-pressure conditions could be detected ahead of the drill there would be significant economic and safety benefit.

Remote detection of such features may involve acoustic or electromagnetic probing. Such advances would have a rapid and profound impact on the combined drilling industries.

Stress, Rock Strength, and Pore Pressure

Stress, rock strength, and pore pressure help determine the degree of difficulty that a given bit at a given drilling state will encounter. These formation attributes are also key in the detection of overpressuring and possible blowout risks. Currently, estimates of the state of stress and rock strength are often difficult and expensive to obtain, and may require complicated procedures such as fracturing the formation or slow leak-off tests (Kunze and Steiger, 1992). A promising new tool to measure borehole deformation and fracture under pressure has recently been developed (Despax and others, 1989; Kuhlman and others, 1993); this development is regarded as an area of great potential for innovative new measurements and analysis.

Drill Pipe-Rock Interaction

Drill pipe interaction with the rock is an important area for evaluation. Drilling loses efficiency as energy is spent in friction with the rock. In addition, drill pipe-rock interaction decreases wellbore stability, resulting in reduced efficiency and, at worst, a stuck drill pipe or the need to abandon the hole (Steiger and Leung, 1990; Gibson and Tayler, 1992). Currently, no sensing system exists to identify depths and friction magnitudes. The state of stress, which is also not currently measured, is one of the most important parameters governing wellbore stability (Hill and others, 1992). For example, if it were known that the friction was concentrated at a depth of weak and friable shale and that a stronger unit lay slightly deeper, the drilling system parameters might be altered to minimize wellbore friction until the weak lithology was passed.

Position and Direction Analysis

Oil and gas wells are increasingly being drilled horizontally or at great inclinations from vertical. Horizontal drilling also is prevalent in the field of environmental remediation (Kaback and others, 1989). Although horizontal drilling is more expensive than conventional vertical drilling, it has the advantage of a greater length of the wellbore being in direct contact with the formation. For environmental remediation applications, this means that, in general, a greater fraction of a subsurface volume can be reached for remediation. In oil and gas wells, the length of the wellbore adjacent to quantities of oil and/or gas is dramatically increased, often by as much as a factor of 100.

High-resolution, three-dimensional seismic surveys help to integrate directional drilling and the subsurface visualization that is provided by seismic data. Horizontal and directional drilling requires accuracy in the determination of bit location and in visualization of the drilling direction in three dimensions. Although some sensing systems exist in this domain (Gaudio and Beasley, 1991; Stephenson and Wilson, 1992; Tarr and others, 1992), major advances in directional drilling will require increased accuracy in these systems.

Cuttings and Mud Analysis

In terms of its composition and properties, the mud column (i.e., the vertical column of drilling mud in the borehole) is a dynamic system whose characteristics are frequently changing dramatically in both time and space. The mud composition changes as shales slough into the column and are dispersed into the mud, or by chemical interaction between the mud and the formation.

Mud fluids are commonly filtered out of the column by the formation of a mudcake along the borehole. These mud fluids flow into the pore space and may interact with minerals in the formation. Compositional changes and varying temperature conditions in the mud column alter its rheological properties; these changes are not monitored. Sensing systems that monitor mud composition and rheological properties downhole, as well as at the surface, could lead to intelligent decisions about remedial mud modification and thus to more efficient drilling.

Environmental Logging

A new area of drilling advancement concerns logging in wells that are drilled for environmental evaluation or remediation. Environmental logging and measurement can identify and quantify hazardous materials, such as radioactive materials and dense nonaqueous phase liquids. Conventional logging has focused on measurements most applicable to oil and gas exploration and development. Although some of these techniques may be applicable to environmental logging, it is likely that significantly advanced systems for detection and monitoring will be needed that can respond to contaminant concentrations at levels of parts per million or parts per billion. For example, high-purity germanium gamma-ray detectors and some of the newer rare-earth orthosilicate detectors might be readily adaptable for radioactive element detection and monitoring. Fluid and rock sampling for uphole analysis is another need in environmental remediation. The potential exists for interactions among the remediation industry, the logging industry, and the national laboratories for applied research in this area.

Environmental remediation itself can benefit from advances in drilling (Kaback and others, 1989). Novel drilling techniques, such as short turning radius, might be combined with innovative remediation systems to improve the efficiency of current practices dramatically.

Telemetry

A mud-pulse telemetry system is used to convey to the surface, measurements of the bit-rock interaction, formation properties, state of stress, and direction and orientation of the drill (Figure 6.3). Because coaxial cables or other data transmission devices are not viable in today's drilling systems, only the most vital information processed downhole can

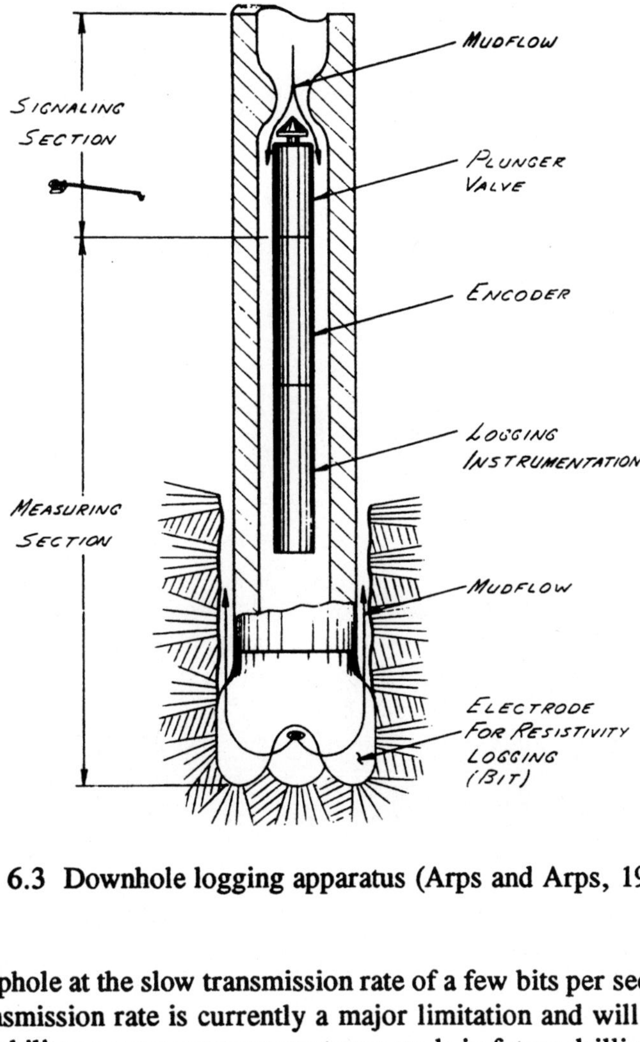

FIGURE 6.3 Downhole logging apparatus (Arps and Arps, 1964).

be sent uphole at the slow transmission rate of a few bits per second. This slow transmission rate is currently a major limitation and will become a greater liability as more measurements are made in future drilling systems.

Data Analysis and Activation Systems

Measurements made downhole or in tunneling, mining, and environmental systems require action by the driller. For example, the trajectory may need modification, the intended target may have been reached, or a different trajectory may be needed until new data dictate otherwise. This process of data analysis is currently performed uphole by humans. The analysis involves comparison of data with external standards, as well as comparison of data generated by a forward model of what is expected according to preexisting concepts of the subsurface. It is possible that much of the decision-making capabilities might eventually reside in a downhole expert system. This would alleviate the need for increased data transmission to an uphole operator.

Priorities for R&D

R&D is needed to produce advanced, intelligent drilling systems that incorporate advanced sensors and microprocessors. The committee has identified eight specific areas for development:

1. *Bit-rock interaction*: A greater understanding of the fundamental physics and chemistry of the bit-rock interaction will aid in advancing downhole sensing capabilities. Two particular areas of research that can be accomplished in the short term are recommended: (a) development of devices to sense the condition of the bit in order to warn of excessive wear or the need for replacement; and (b) development of sensors for the mud properties at the bit and throughout the wellbore to improve borehole quality and mud conditioning. Longer-term research should seek to develop a means of sensing the rock breakage mechanism to aid in bit selection strategies.

2. *Formation evaluation and fracture detection at and ahead of the bit*: Although significant advances in MWD technologies have occurred during the last decade, great need remains for improved capabilities to make measurements around and ahead of the bit. Three long-term research areas are recommended: (a) development of fracture-detection capabilities based on electromagnetic, acoustic, and other innovative methods such as instantaneous mud loss; (b) development of advanced formation evaluation procedures based on sensor measurements; and (c) development of high-resolution imaging capabilities to aid in formation evaluation and steering decisions.

3. *Stress, rock strength, and pore pressure*: Advances in sensing technologies will lead to an improved ability to measure in situ stress, rock

strength, and pore pressure. Two long-term research efforts in this area are recommended: (a) the development of novel direct sensing of stress, rock strength, and pore pressure for optimizing drilling parameters, steering, minimizing formation damage, and formation evaluation; and (b) the development of technologies to predict pore pressure from acoustic measurements.

4. *Drill pipe-rock interaction*: An improved understanding and means of evaluating drill pipe interaction with rock can lead to greater drilling efficiencies and increased wellbore stability. Two short-term research areas are recommended: (a) the development of sensors to detect, locate, and measure friction between the drill string and the formation to optimize drilling parameters and enhance wellbore stability; and (b) the development of sensors to determine drill pipe wear.

5. *Position and direction analysis*: Advances in directional drilling are tied to improved capabilities for determining bit location and position. Two long-term research efforts in this area are recommended: (a) the development of sensing systems capable of determining the bit position (depth and spatial coordinates) within an accuracy of 1 ft; and (b) the development of enhanced systems to determine bit direction and the resultant forces to optimize steering capabilities.

6. *Environmental logging*: Environmental evaluation and remediation would benefit from improvements in sensing systems for drilling, especially detection and monitoring of contaminants, fluid and rock sampling, and novel drilling techniques for remediation purposes. In the short term, research should be supported to develop advanced methods of sampling Earth materials. Longer-term research needs include (a) the development of advanced sensors to detect environmentally hazardous materials to aid in remediation efforts; and (b) the development of innovative techniques and sensors to monitor contaminated sites on a long-term basis.

7. *Telemetry*: Drilling data generated and recorded downhole need to be transmitted uphole in a speedy and efficient manner. Present telemetry systems are capable of transmitting data at rates of a few bits per second. The incorporation of advanced sensors into the drilling system requires much higher rates of data transmission. Consequently, telemetry

systems capable of transmitting data at rates of kilobits per second should be developed.

8. *Data analysis and activation systems*: Downhole data analysis and activation systems may improve overall drilling operations by reducing the need for data transmission uphole for operator decision making. Long-term research is recommended to develop sophisticated software for downhole analysis of drilling data.

References

Arps, J. J., and Arps, J. L., 1964, The subsurface telemetry problem—a practical solution: Journal of Petroleum Technology, v. 16 (5), p. 487-493.

Betts, P., Bliunt, C., Broman, B., Clark, Hibbard, L., Louis, A., and Oosthoek, P., 1990, Acquiring and interpreting logs in horizontal wells: Oilfield Review, v. 2 (3), p. 34-51.

Cantrell, L. A., Paxson, K. B., Keyser, B. L., and Ball, S., 1992, Case histories of MWD as wireline replacement: An evolution of formation evaluation philosophy: Proceedings SPE Annual Technical Conference, Washington, D.C., Oct. 4-7, p. 115-130.

Desbrandes, R., 1982, Diagraphies dans les Sondages: Paris, Editions Technip, 584 pp.

Despax, D. M., Martin, J. P. A., and Gueret, F. C., 1989, Geomechanical Probe for a Drilling Well: U.S. Patent 4,800,753, 11 pp.

Doveton, J. H., 1986, Log Analysis of Subsurface Geology: New York, John Wiley and Sons, 273 pp.

Ellis, D. V., 1987, Well Logging for Earth Scientists: New York, Elsevier, 532 pp.

Fay, H., 1993, Practical evaluation of rock-bit wear during drilling: SPE Drilling and Completion, v. 8 (2), p. 99-104.

Gaudio, G. B., and Beasley, J. C., 1991, A comparison of MWD and wireline steering tool guidance systems in horizontal drilling: Proceedings SPE Annual Technical Conference, Dallas, Tex., Oct. 6-9, Richardson, Tex., p. 7-18.

Gianzero, S., Chemali, R., Lin, Y., Su, S., and Foster, M., 1985, A new resistivity tool for measurement-while-drilling: Society of Professional

Well Log Analysts 26th Annual Logging Symposium Proceedings, p. A1-22.

Gibson, M. T., and Tayler, P. J., 1992, Optimal applications engineering and borehole stability analysis avoid differential sticking and lead to successful open hole completion of North Sea horizontal well: Proceedings SPE Annual Technical Conference, Washington, D.C., Oct. 4-7, p. 741-754.

Hill, T. H., Seshadri, P. V., and Durhan, K. S., 1992, A unified approach to drillstem-failure prevention: SPE Drilling and Completion, v. 7 (4), p. 254-260.

Jan, Y.-M., and Harrel, J. W., 1987, MWD directional-focused gamma ray—a new tool for formation evaluation and drillng control in horizontal wells: Society of Professional Well Log Analysts 28th Annual Logging Symposium Proceedings, p. A1-A17.

Kaback, D. S., Looney, B. B., Corey, J. C., Wright, L. M., and Steele, J. L., 1989, Horizontal wells for in situ remediation of groundwater and soils: National Water Well Association Outdoor Action Conference, Orlando, Fla., May 22-25, 1989, 16 pp.

Kuhlman, R. D., Heemstra, T. R., Ray, T. G., Lin, P., and Charlez, P. A., 1993, Field tests of downhole extensometer used to obtain formation in-situ stress data: SPE 25905, SPE Rocky Mountain Regional/Low Permeability Reservoirs Symposium, Denver, April 12-14, 1993, Richardson, Tex., SPE, p. 625-634.

Kunze, K. R., and Steiger, R. P., 1992, Accurate in situ stress measurements during drilling operations: Proceedings SPE Annual Technical Conference, Washington, D.C., Oct. 4-7, p. 491-499.

Norve, K. N., and Saether, H., 1989, Field experience using the full suite MWD-combination for reservoir logging and evaluation: Society of Professional Well Log Analysts 30th Annual Logging Symposium Proceedings, p. A1-A26.

Onoe, Y., Yoder, D., and Lawrence, M., 1991, Improved drilling performance, efficiency, and operations using an advanced real-time information system for drilling: Proceedings SPE Annual Technical Conference, Dallas, Oct. 6-9, p. 335-344.

Roy, S., and Cooper, G., 1993, Prevention of bit balling in shales—preliminary results: SPE Drilling and Completion, v. 8 (3), p. 195-200.

Steiger, R. P., and Leung, P. K., 1990, Prediction of wellbore stability in shale formations at great depth: Proceedings ISRM International

Symposium on Rock at Great Depth, Rotterdam, Balkema, v. 3, p. 1209-1218.

Stephenson, M. A., and Wilson, H., 1992, Improving quality control of directional survey with continuous inertial navigation: SPE Drilling and Completion, v. 7 (2), p. 100-106.

Suau, J., Grimaldi, P., Poupon, A., and Sounaite, P., 1972, The Dual Laterolog-R_{xo} Tool: SPE 4018, Society of Petroleum Engineers 47th Annual Fall Meeting, San Antonio, Tex., Oct. 8-11, Richardson, Tex., SPE, v. 21, p. 121-132.

Tarr, B. A., Kuckes, A. F., and Ac, M. V., 1992, Use of new ranging tool to position a vertical well adjacent to an horizontal well: SPE Drilling and Completion, v. 7 (2), p. 93-99.

Tittman, J., 1986, Geophysical Well Logging: Orlando, Academic Press, 175 pp.

Wraight, P., Evans, M., Marienbach, E., Rhein-Knudsen, E., and Best, D., 1989, Combination format in density and neutron porosity measurements while drilling: Society of Professional Well Log Analysts 30th Annual Logging Symposium Proceedings, p. B1-B23.

7
DIRECTIONAL DRILLING AND TUNNELING

Introduction

Tunneling and drilling operations fall into two broad categories: (1) operations in which excavation is carried out by attacking the rock face selectively, and (2) operations that use a device to advance the whole cross section of the excavation simultaneously. The first category includes all drill-and-blast and manual methods, and flexible mining machines such as roadheaders or continuous miners. These methods allow a tunnel of any cross section or any curvature to be driven, provided the machine and its mucking system can negotiate the curve. They are typically used in mining operations where great flexibility in size, cross section, and curvature of the excavation is needed.

The second category consists of excavations in which the whole face is advanced simultaneously, and includes those driven by tunnel boring machines (TBMs), raise borers, and all small-diameter holes drilled by the use of a drill bit. Such holes are almost always of circular cross section, because the excavation method consists of rotating some form of cutter head against the rock face. The drive mechanism for the cutter head is normally quite bulky and is rigidly attached to a cutter head because of the need to transmit the required power. Such drilling assemblies tend to be long relative to their diameters and tend to fit tightly in the hole. Changing the direction of the hole is then a matter of making the cutter head cut sideways and getting the drive mechanism to follow. This operation, whether for a shallow exploratory core hole for prospecting purposes, a production oil well, or a full-size freeway tunnel, has certain basic requirements, even though the equipment used in each case may be very different.

In each operation, the hole direction is changed either by exerting a sideways force on the rock cutting assembly or by tilting the cutting structure so that its axis is inclined to that of the existing borehole or tunnel. If a sideways force is used, it is obtained by reaction of the drilling assembly against the wall of the hole or tunnel, and the deviation is achieved by the ability of the cutting structure to cut sideways as well as forward (Millheim and Warren, 1978; Millheim and Apostal, 1981).

A drilling assembly that exerts a large sideways force, coupled with a cutting structure having substantial sideways cutting ability, will produce a sharply curved borehole. If the hole curvature is generated by tilting the cutting structure, the degree of turn will be determined by the tilt and the effective positions of contact of the rest of the drilling assembly with the hole wall (Figure 7.1). Conversely, straight holes are drilled by rigid drilling assemblies provided with cutter heads or drill bits that have no sideways cutting teeth and that fit snugly in the hole.

Directional Tunneling

Tunnel boring machines excavate a constant cross section of tunnel (usually circular), and the excavating head is not free to move sideways with respect to the main machine body. In driving a tunnel with a TBM, the body of the machine is held in the tunnel by one or more pairs of grippers. In principle, two steering methods are available, either by exerting a sideways force or by tilting the cutting head. In a typical TBM, there is a set of grippers, guides, or side supports close to the cutting head, which is used to position the head, and another set further back that is used to brace the machine against the reaction from the cutting feed force. The preferred method of steering consists of yawing the axis of the machine by displacing the main beam with respect to the rear gripper unit, with the result that the machine pivots about the forward gripped region. No sideways force is applied to the cutter head, so the machine axis is always tangential to the curve of the tunnel.

To achieve a sharper turn, sideways force can be applied to the cutter head by extending one of the forward side supports so that it bears against the tunnel wall (Robbins Co., 1993). In this way, the cutter head is forced to cut sideways as well as forward (Figure 7.2). In this "crabbing" mode, the machine axis is no longer tangential to the curve, and care must be taken not to damage the machine by forcing it too hard into the side wall.

In the last few years, there has been a rapid development of microtunneling machines, or moles. These are unmanned machines,

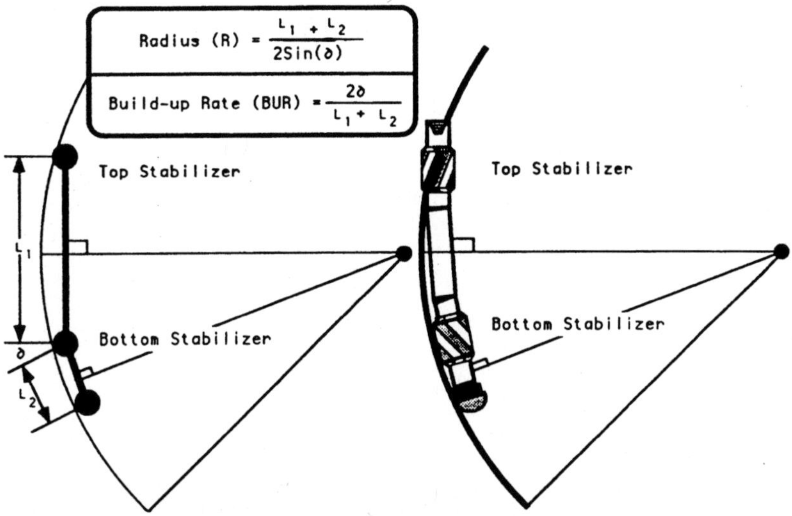

FIGURE 7.1 Hole curvature determined by the position of contact points with the hole wall (Karlsson and others, 1989).

capable of driving tunnels of, typically, 6 to 30 inches in diameter. They have been developed mainly in Japan (Akiba and Yamada, 1989; Hayashi and Miyata, 1989; Moria and Sawaguchi, 1989) and Germany (Beaumont, 1989) for sewer work. The machines consist of a powered cutter head and steering assembly that is pushed forward by a jacking system from a launching pit. As the tunnel advances, sections of temporary or permanent lining are fed in and pushed forward behind the mole by the jacking unit. Maximum tunnel length is up to 500 ft (Hayashi and Miyata, 1989). The excavation method is by pressure/thrust, auger, water pressure balance, slurry shield, or other methods. The machines operate in a variety of relatively soft soils, although some are provided with crushing heads that will cope with gravel and small boulders. Some of the machines are steerable and typically use a system of hydraulic jacks to tilt the cutter head in the required direction.

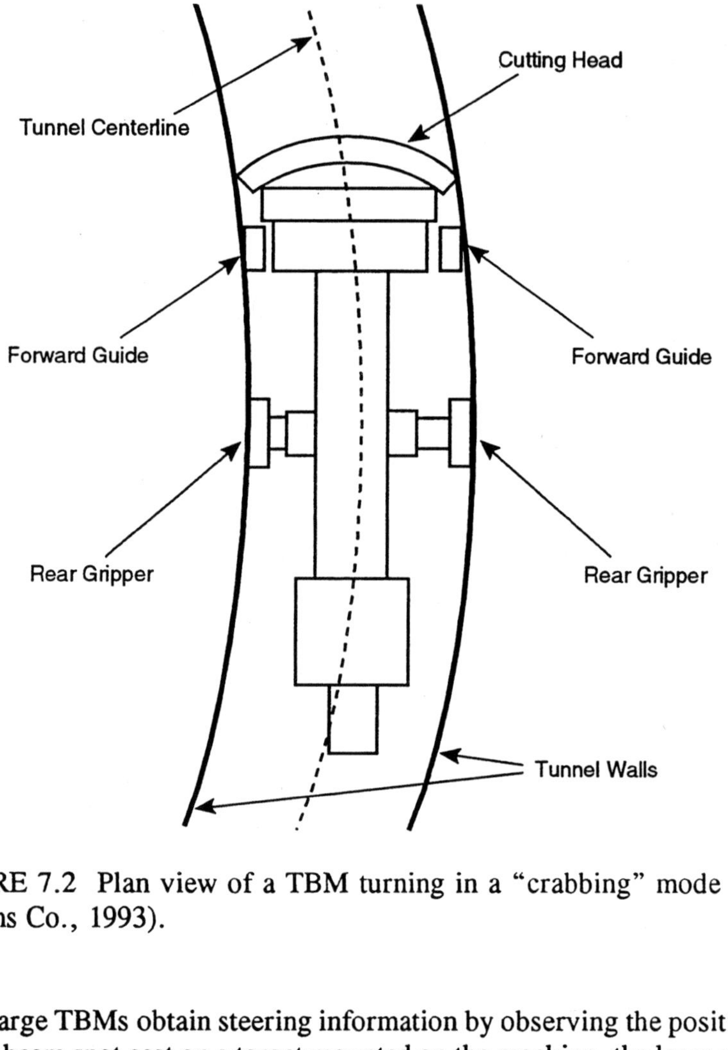

FIGURE 7.2 Plan view of a TBM turning in a "crabbing" mode (after Robbins Co., 1993).

Large TBMs obtain steering information by observing the position of a laser beam spot cast on a target mounted on the machine, the laser source being mounted on a fixed support some way to the rear of the tunnel. Several manufacturers offer such systems (Anon., 1988).

A disadvantage of the laser system is that the laser must be within line of sight of the tunneling machine, and it is therefore not well adapted to surveying tunnels with numerous or sharp curves. At present, steering adjustments to the machine are carried out manually, but automated systems are under development (Gibson, 1993). The Japanese have an alternative system using gyroscopes, also with manual implementation of the steering information (Miyazaki, 1993). For the same reasons, the gyroscopic system seems better adapted to the remotely operated microtunneling machines that drive tunnels a couple of feet or so in diameter. A disadvantage of the gyroscopic system is that it must be reset periodically, for example, by removing it from the tunnel so it can "see" a reference satellite.

Directional Drilling of Boreholes

Reasons for Directional Drilling

Directional drilling (e.g., Cooper, 1994) is not new. Since the earliest times, boreholes were made to deviate by placing tapered wedges or "whipstocks" in the borehole to force the bit sideways into a new direction, and it was known that different bottom-hole assemblies (BHAs) had a tendency either to increase or to decrease the inclination of the hole. Most developments in directional drilling started in the petroleum industry and then spread to other fields. Pioneering work was carried out in oil well drilling by John Zublin, H. John Eastman, and others during the 1950s in drilling directional drainholes from existing wells (Stormont, 1953; Eastman, 1954). Significant work was also done during the 1960s in directional drilling from Huntington Beach, California. However, progress was slow until the mid-1980s, when it was realized that significant increases in productivity could be obtained by orienting the borehole correctly in the reservoir (Reiss, 1987). Simultaneously, it was realized that substantial economies could be obtained in offshore operations if, by extending the horizontal reach of the production wells, the area accessed from any platform could be increased (Wilson and Willis, 1986) or the number of platforms needed to exploit a reservoir could be reduced (Tolle and Dellinger, 1986).

Since the mid-1980s, a large amount of work has been done to investigate all aspects of directional drilling. In addition to the control of direction, other important issues include surveying methods; borehole stability and cuttings removal in inclined boreholes; the reservoir engineering aspects of horizontal wells; and the various special problems associated with completion and production, such as the cementing of liners, gravel packing, and fracturing.

Directionally drilled wells fall into three main categories. In the first category, the task is to reach locations that are not accessible through straight, vertical holes. For example, unfavorable surface topography may exist, such as the presence of buildings, hills, or water above the target location. Alternatively, a directional well may be drilled to avoid undesirable locations in the subsurface, including fault zones or regions of known "difficult" rock (e.g., salt, swelling shales, and high-pressure regions). It may be advisable to drill a directional well from a safe distance to intersect a blown-out well. In exploration drilling, several deviated test bores may be drilled from the same upper section of the hole in order to save on drilling costs and avoid moving the rig.

In the second category, the objective is to reach a substantial distance horizontally away from the drilling location. This technique, known as "extended reach" drilling, is used to allow many parts of the reservoir to be accessed from one location. The major application of extended reach drilling is in offshore operations, where it is used to reduce the number of platforms needed. Recently, extended reach drilling has become of interest for land operations, where environmental concerns or urban space restrictions may impose land use demands. An area of growing interest for extended reach drilling is to access offshore oil fields from a shore-based location, resulting in reduced environmental impact and reduced cost associated with land-based drilling.

In extended reach drilling, maximum reach depends on the target depth, as well as maintenance of good borehole conditions and a sufficiently steep angle to allow the drilling assembly to slide forward so as to apply weight to the bit. The technology is advancing rapidly. The current world record extended reach well, drilled in the North Sea, accessed a target at a horizontal distance of 23,917 ft (4.53 miles) at a depth of about 9,000 ft (Anon., 1993b). The length of the hole was 28,743 ft (5.44 miles).

The third category consists of wells in which the part of the well that lies in the reservoir is given a particular orientation so as to increase productivity. For example, in a vertically thin reservoir, a horizontal hole can contact a greater part of the reservoir than a vertical one, increasing the drainage contact area and delaying the water or gas coning. The reservoir geometry helps determine whether to use a horizontal or vertical well; a horizontal well is more productive and thus more favorable than a vertical well as the reservoir width increases and the height decreases.

Actual values have been estimated by Joshi (1987a, b, 1988; Figure 7.3). For example, for the same density of wells per unit of land surface (30-acre spacing), a 1,000-ft-long horizontal well has about twice the productivity of a vertical well if the reservoir thickness is 400 ft, and this figure increases about five times if the reservoir height decreases to 100 ft. When the reservoir thickness is 25 ft, the productivity becomes seven times that of a vertical well. In addition, drilling a horizontal hole in a tight but fractured reservoir so as to intersect the maximum number of fractures, which are usually vertical, can greatly increase productivity (Reiss and others, 1984; Reiss, 1987; Haas and Stokley, 1989).

In some circumstances, one or more wells are drilled horizontally into the producing formation from a main vertical well. These are often referred to as "drainholes," whereas a single borehole drilled horizontally

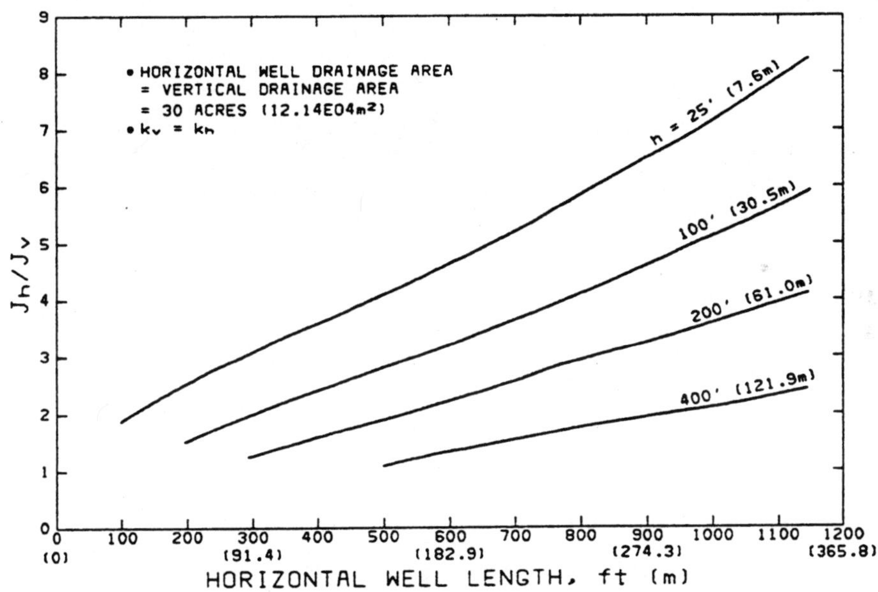

FIGURE 7.3 The ratio of productivity indices for horizontal and vertical wells for different horizontal well lengths and reservoir heights (Joshi, 1987a).

into the producing formation is referred to as a horizontal well. As with extended reach drilling, the technology is advancing rapidly. Texaco recently claimed a record for a pair of horizontal drainholes that extended 5,332 and 4,100 ft from a single vertical well (Anon., 1993c).

Limitations to Borehole Reach and Turning Radius

In drilling any hole, it is essential that sufficient thrust be transmitted to the bit to make it advance. In an extended reach hole, this requirement becomes one of making the contact between the drilling assembly and the hole wall as friction free as possible, allowing for the flattest "sail angle." (For the record-breaking hole recently drilled in the North Sea, this was 82 degrees from the vertical; Anon., 1993b.) For hole angles flatter than this and for horizontal holes, the drilling assembly has to be pushed forward by the action of a thrusting section of heavy pipe or collars located further up the hole in a more steeply inclined interval (Figure 7.4).

Drill pipe located between the thrusting section and the drill bit must then operate in compression and is subject to buckling (Lubinski, 1950). In an inclined hole, the buckling is not as severe as it would be in a vertical hole. This is because of the stabilizing effect resulting from the pipe being held by gravity in the trough formed by the lower side of the hole (Dawson and Paslay, 1984). Buckling forces the pipe against the hole wall and thus reduces the efficiency with which the thrust force can be transmitted. In addition, buckling (and bending the drill string to negotiate the curve) induces flexural fatigue, particularly if the pipe is made to rotate, as is required if the drill bit is rotated from surface.

To strengthen the pipe against buckling, it is made thicker and often is provided with "wear knots" of wear-resistant alloys (Figure 7.4). However, the thicker the pipe, the greater is its weight, and the greater is the drag force it exerts on the hole wall. Furthermore, the thicker the pipe, the smaller is the curvature through which it can be bent before the axial stress in the pipe reaches the fatigue limit. These considerations have led some to suggest the use of aluminum, which has a lower modulus and a lower density (Glagola and Wong, 1986); however, its use is not yet widespread.

It is clear that different compromises must be accepted in combining reach with hole curvature, and several different engineering solutions have

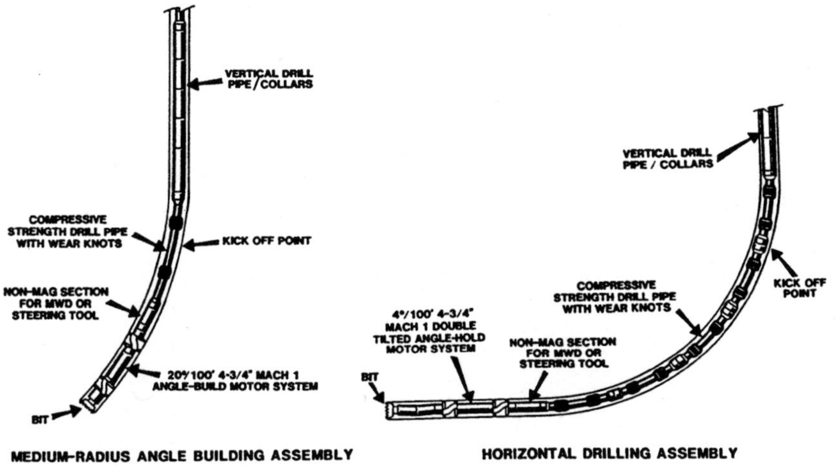

FIGURE 7.4 Typical arrangements for horizontal and directional drilling (Logan, 1988).

been developed to fill particular requirements. In general, holes that have sharp radii of curvature are drilled at a smaller diameter and achieve less horizontal displacement. Drilling methods for small-radius curves seldom entail rotating the bit from the surface, and use downhole motors or unconventional methods for breaking the rock. Holes with large radii of curvature reach further and may often be drilled by rotation from the surface by using conventional full-size drilling tubulars.

Several considerations govern the choice of curve radius and therefore the drilling method. For extended reach wells, the maximum reach is attained with large tubulars and gentle curvature. However, a hole of gentle curvature necessarily has a long curved section that must start a substantial distance above and horizontally displaced from the target. This may not be convenient for several reasons, including the following:

1. *Precision of response in steering*: A hole with gentle curvature will have a long curved section. The curve may intersect several strata on the way to the target, each causing a different turning response in the

BHA. Correction of the trajectory will take a long distance, so it may be difficult to have information early enough to direct the well accurately to the target.

2. *Uncertainty of target position*: In many cases, the exact depth of the target is uncertain, and decisions concerning the steering of a large-radius curve may have to be made hundreds of feet above the target before its position is clearly known.

3. *Avoidance of geologic difficulties*: Another disadvantage of a long curved section is that substantial footage will be drilled at high inclination in nonreservoir rock. This is likely to be the reservoir caprock, such as a weak or water-sensitive shale (Hourcard and Bannerman, 1990). Reducing the turning radius of the hole will reduce the distance to be drilled in the problem strata and, because the kick-off point will be lowered, may allow some strata to be avoided entirely by the curved section.

For the above reasons, holes are often drilled with two curved sections joined by an inclined straight interval (Schuh, 1989; Figure 7.5). This allows the turning tendency of the BHA to be assessed during the drilling of the first curve and the straight section to be shortened or lengthened to allow the second curve to hit the target exactly. In addition, the straight section may be continued down to intersect the target horizon at a steep angle (Austin and others, 1988; Stacey and others, 1992). This will allow the driller to determine its exact depth, and perhaps to recover cores or log the well. Once this information is known, the well may be plugged back and the second curve drilled with high precision to target. For wells drilled with sharp curvature, the vertical section of the well may itself be continued into the target and the kick-off points determined after the productive interval has been located and logged (Austin and others, 1988; White, 1989).

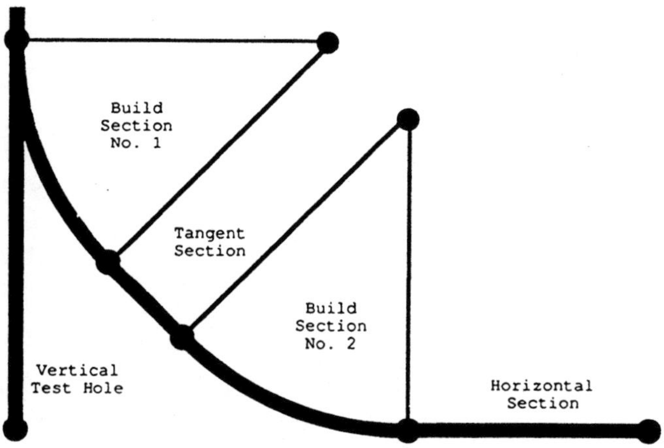

FIGURE 7.5 Well profile showing two build sections with a straight portion between, and the initial test hole to determine reservoir depth (Karlsson and others, 1989).

Methods for Directional Drilling

Early techniques for directional drilling used whipstocks of various types to force the bit sideways. In soft ground, a bit with one oversized nozzle was used to erode one side of the hole preferentially (while not rotating) to achieve the same effect. A borehole may have its inclination changed in the vertical plane (but not in azimuth)[1] by the judicious placement of stabilizers behind the bit. A stabilizer placed close to the bit, followed by another a substantial distance away will allow the section of collar between them to sag toward the bottom of the hole. This will tip the bit upward and result in the bit drilling so as to increase the deviation from vertical. This is known as "building angle." Conversely, a first stabilizer located a distance behind the bit will allow the bit to hang down and drill

[1] Some bits, particularly polycrystalline diamond cutter (PDC) drag bits, do tend to change azimuth by "walking" to the left or right. The response varies from bit to bit and according to the lithology.

toward the vertical. This is known as "dropping angle." Figure 7.6 shows sketches of these configurations for a flexible BHA.

Such "pendulum assemblies" rely on the borehole having some initial inclination, and are simple to use in terrain for which the building and dropping characteristics of each assembly are known. These methods rely both on tilting the bit face in the new direction and on forcing it sideways, and therefore combine the two fundamental methods for changing direction. The degree of curvature that may be achieved is quite low, typically 1 to 5 degrees per hundred feet, and the use of pendulum assemblies is generally restricted to full-size (greater than 6 inches) drilling assemblies.

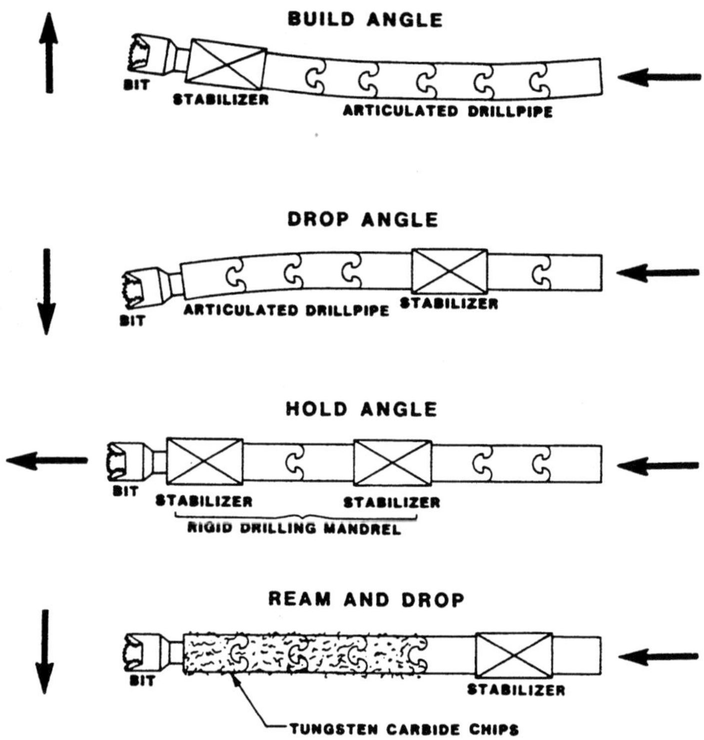

FIGURE 7.6 Arrangement of stabilizers to build, drop, or hold angle (Logan and others, 1987).

To obtain greater curvature, gain better control of the turning radius, and be able to change the azimuth of the hole, several specific engineering solutions have been developed. Because of the compromises in balancing turning radius against reach and the resultant implications for borehole diameter discussed above, no one drilling method is satisfactory for all radii of curvature. It is customary to distinguish among these as *long-, medium-, short-, and ultra-short-radius* methods (Figure 7.7).

Long-Radius Methods

Long-radius methods use conventional drilling tubulars and are applied in hole diameters 6 inches or larger (Bosio and others, 1987). Curvatures of 2 to 5 degrees per hundred feet are typical, resulting in turning radii of 300 to 1,000 ft or so. Pendulum angle building or dropping

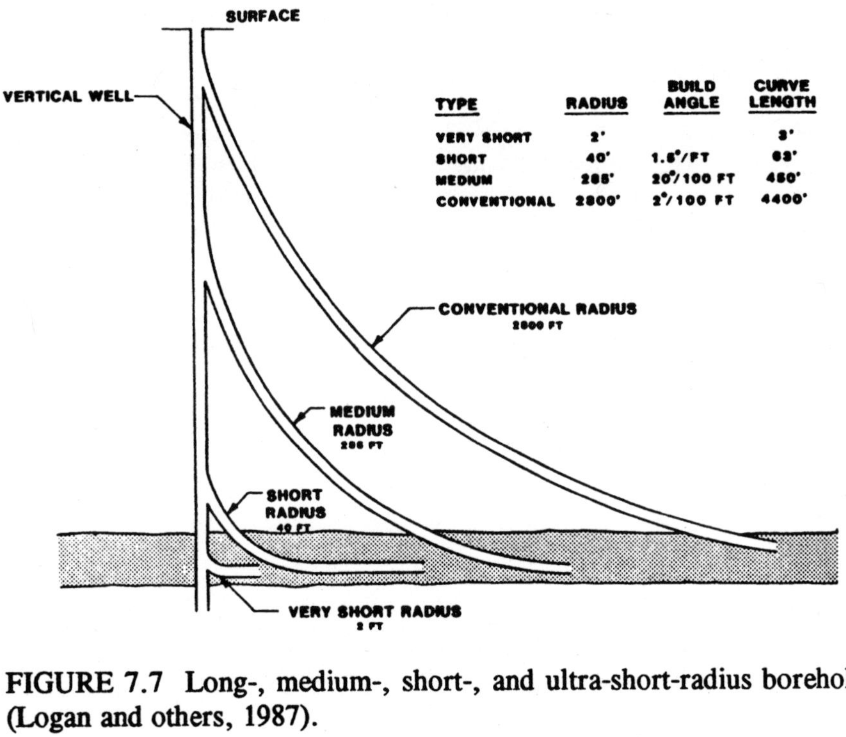

FIGURE 7.7 Long-, medium-, short-, and ultra-short-radius boreholes (Logan and others, 1987).

assemblies may be used to change the inclination of an already inclined hole. To change azimuth or to gain more precise control on the initial kickoff, drilling is carried out with a combination of downhole motor and "bent sub." The latter is a section of drill collar that has a bend of a few degrees between the top and bottom connections, which both forces sideways and tilts the motor section located below it. To change the direction of the hole, pipe rotation is stopped, and the drill pipe position is adjusted from the surface until the bent sub is oriented so that the "tool-face orientation" (i.e., the angle between the projection of the bend axis and the high side of the hole on the hole bottom) is set correctly. The motor is then started, and drilling is allowed to proceed. Periodic measurements are made of hole direction and inclination, with adjustments to the tool face angle as necessary. If it is desired to drill straight ahead, this may be achieved without removing the directional drilling assembly, by applying a slow rotation (some tens of revolutions per minute) to the drill stem from the surface. The motor and bent sub will continue to drill while rotating, in effect drilling a very fine spiral that is indistinguishable from a straight hole.

Different motors are available. For drilling with mud, turbines or positive displacement motors (PDMs) are available. The former were the first to be introduced, but produce their best power output at rotary speeds that are too high for most present bits (with the exception of some diamond bits). PDMs have undergone substantial development (Makohl and Jurgens, 1986) and are now generally preferred. For air drilling, downhole hammers are commonly used (Whiteley and England, 1985), and recently there have been substantial developments in PDMs for air drilling (Shale, 1991; see also Lattimore, and others, 1987; Yost and others, 1987).

It should be realized that drilling with a motor and bent sub costs more than drilling with a pendulum assembly, which is preferred if no changes in azimuth are needed. The response of a pendulum assembly is sometimes difficult to predict and changes as the lithology changes. Recently, however, stabilizers have been introduced that are adjustable from the surface, thereby allowing changes to be made to the building or dropping tendency without having to remove the BHA from the hole (Eddison and Symons, 1990).

The great advantage of drilling with gentle curvature is that the full range of drilling and completion equipment can negotiate the curve. Not only does this allow all logging and other tools to reach the bottom of the well, but the well may be cased along its whole length. This allows selective completion of particular intervals, as well as the control of hole collapse, sand production, and other problems.

Medium-Radius Methods

Medium-radius methods use drilling assemblies that are capable of rates of turn of 5 to 30 degrees per hundred feet, resulting in radii upward of a few hundred feet (20 degrees per 100 ft gives a radius of 287 ft). The dividing line between long- and medium-radius methods is not precise, and is best thought of in terms of the completion practice; a long-radius well can normally be drilled and completed with any of the conventional casing and completion methods, whereas a medium-radius well has too sharp a curvature to accommodate many standard completion tubulars. A medium-radius well will frequently be drilled with an open hole section at the end or will require special treatment. Since the maximum curvature allowable in a casing depends on its diameter, the distinction between long- and medium-radius wells will therefore be somewhat dependent on hole diameter.

Many manufacturers offer drilling assemblies that are capable of drilling medium-radius wells. These are bent sub/motor combinations that frequently involve the use of adjustable bend devices so that the degree of bend, and hence the rate of turn, can be adjusted in the field. The adjustment may be made by altering a single bend or by using various types of double bend that can be combined to enhance or cancel each other so as to generate different rates of turn (Figure 7.8). For examples of the use of these different systems, see Edlund (1987), Barrett and Lyon (1988), Karlsson and others (1989), Sheikholeslami and others (1989) for Hughes Christensen equipment; Rehm and Garcia (1989) for Becfield; Reiss and others (1984) for Telepilote; and Anon. (1989) for a description of several systems.

An interesting feature of the Hughes Christensen "Navigation Drilling System" is that whereas most of the bent sub/motor combinations have the motor axis displaced from the main axis of the drilling assembly, and therefore push the bit sideways as well as tilting it, the Navigation Drilling System uses a double bend to tilt the bit from below the motor, thus causing the deviation of the hole to occur mostly by the tilt of the bit. This is claimed to give more consistent steering.

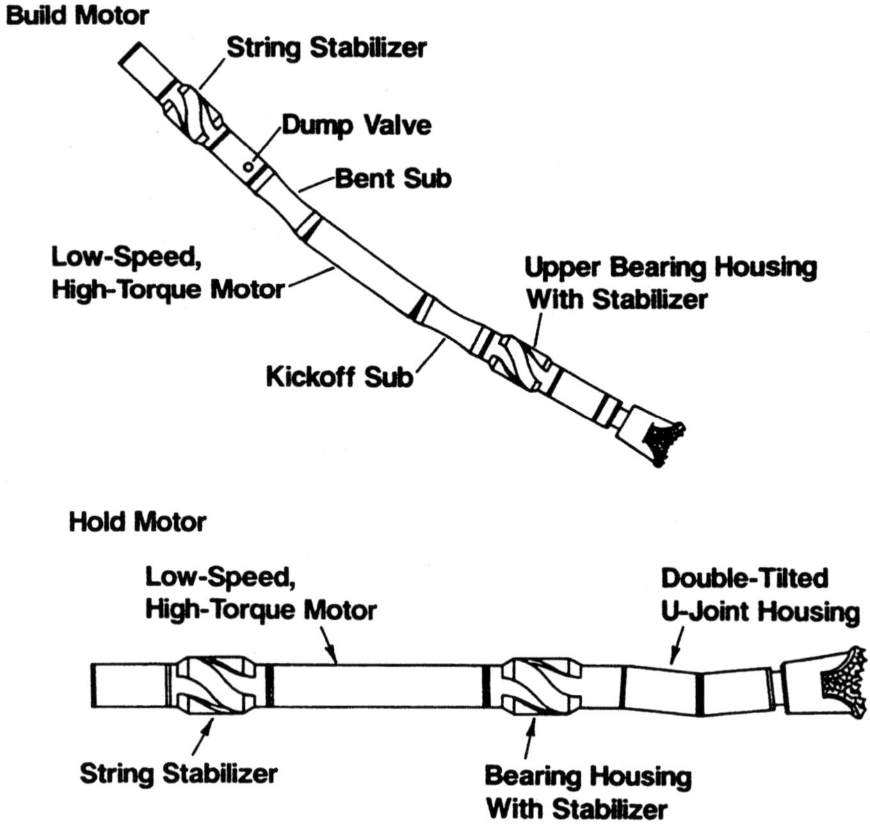

FIGURE 7.8 Equipment for drilling medium-radius curves (Sheikholeslami and others, 1989).

Short-Radius Methods

A short-radius drilling method described by Parsons and Fincher (1986) is capable of rates of turn of 1.5 to 3 degrees per foot, resulting in the borehole becoming horizontal in 30 to 40 ft of drilled hole for a system operating with a 4.5-inch bit, or in 50 to 60 ft for the 6.25-inch-bit system.

The equipment consists of an orientable whipstock that is lowered into a previously drilled vertical hole, a directional drilling assembly, and a straight drilling assembly (Figure 7.9).

After the whipstock has been placed and oriented in the hole, the directional drilling assembly is lowered. This consists of a nonrotating external tubular guide, through which is passed a flexible drive tube supported on bearings at either end. The guide has slots milled in it to cut about two-thirds of its circumference and is then bent so that it forms a springy curve of 20- to 40-ft radius, depending on the guide diameter. This guide is sufficiently flexible to be forced straight to enter the vertical hole, but in this state it exerts about 1,000 lb of sideways force, so that upon leaving the whipstock, the assembly drills ahead along a curve. In front of the guide is a conventional bit, while to the rear of the guide are lengths of flexible drive pipe, composed of interlocking steel elements that allow torque to be transmitted even though the pipe can flex. (The flexible pipe has a polymeric liner to make it leakproof, so as to carry the drilling mud.)

In operation, the directional drilling assembly is made to drill ahead until the desired degree of inclination is achieved. It is then removed and replaced by a straight drilling assembly that is allowed to drill forward until the desired length of hole is drilled, usually 200 to 400 ft (maximum 900 ft). A clutch arrangement allows the orientation of the guide to be adjusted initially, but the azimuth of the hole cannot be changed easily once it is started because the flexing of the guide is limited to one plane.

The accuracy of the azimuth is typically about $\pm$ 20 degrees, although when drilling multiple wells, experience allows better predictive capability. Inclination can, of course, be controlled by drilling more or less of the curve. Additional equipment for coring is available (Eaton, 1990). This method has been used successfully in several applications in the oil industry (Keelean and others, 1989) and for coalbed methane recovery (Logan and others, 1987).

Ultra-Short-Radius Methods

Ultra-short-radius drilling methods capable of producing horizontal drainholes directly from an underreamed length of a vertical wellbore have been described by Dickinson and Dickinson (1985) and by Dickinson and others (1989). An erectable whipstock placed in a 22-inch-diameter underreamed section of a 7-inch diameter vertical hole turns a length of

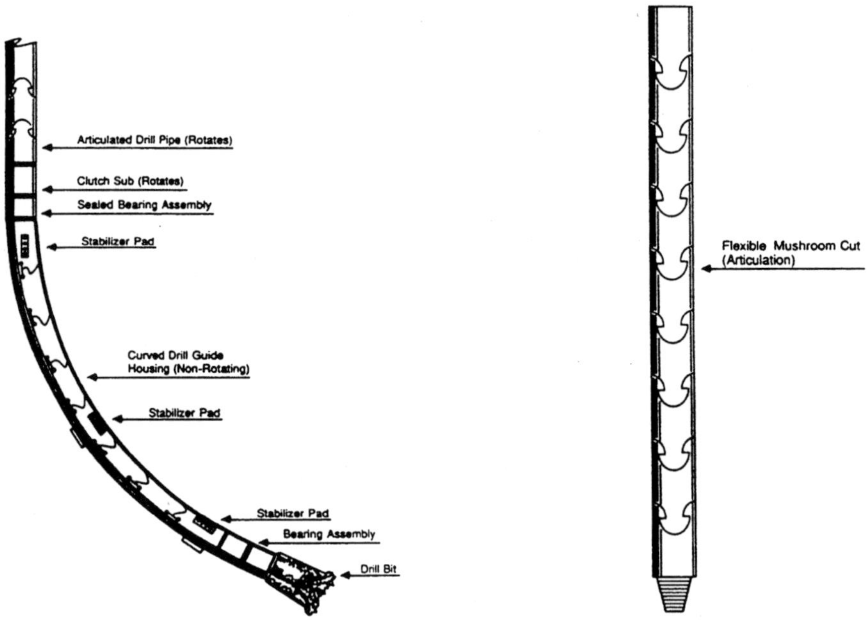

FIGURE 7.9 Equipment for drilling short-radius curves: left, the guide; right, articulated drill pipe (Keelean and others, 1989).

1.25-inch-diameter electric resistance welded tubing from vertical to horizontal with a radius of 1 to 2 ft. At the forward end of the tubing, a jet nozzle allows high-pressure water to be directed at the rock to be drilled, while the pressure of the water inside the tubing feeds it forward into the hole created by the water jet. A horizontal drainhole 1.5 to 2.5 inches in diameter is thus created, which, under favorable circumstances, extends a hundred or more feet from the vertical well (Dickinson and others, 1992; Figure 7.10).

After the hole is drilled, it may be gravel packed, and the tubing may be perforated and severed electrochemically (Dickinson and others, 1987). This allows the whipstock to be turned and/or lifted so that more radial holes may be drilled. A means for steering the tubing as it exits the whipstock, described by Dickinson and others (1992), consists of a sensing device and equipment for allowing vertical deviation control by selectively opening side ports near the forward end of the tubing. Side forces generat-

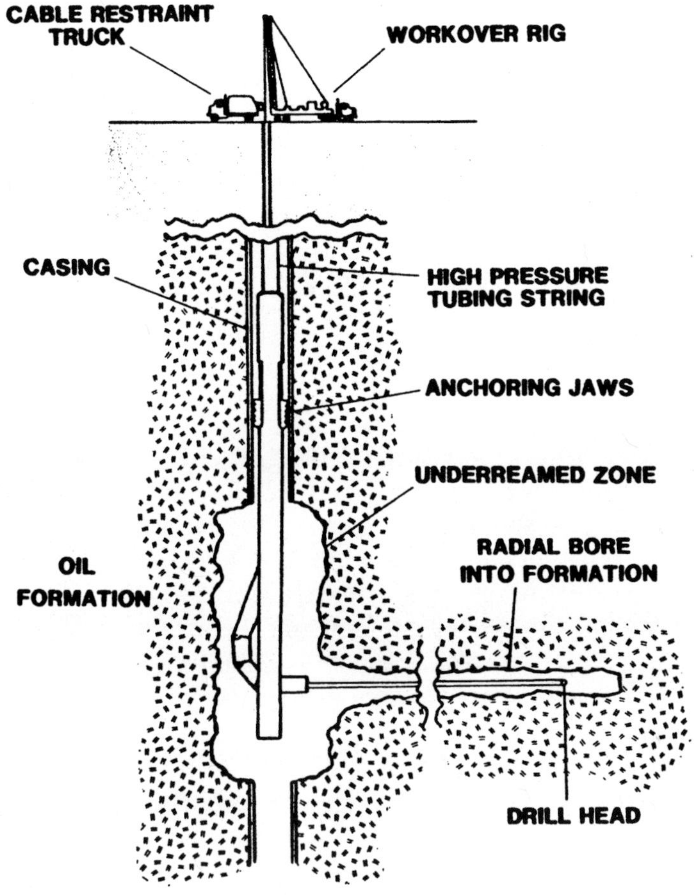

FIGURE 7.10 Ultra-short-radius system (Dickinson and others, 1987).

ed by the exit of high-pressure water from the ports force the drilling head up or down (Figure 7.11). This steering method is of interest because it does not generate the steering force by pressing some part of the drilling assembly against the hole wall. This steering method has also been used elsewhere (Ritchie and others, 1989).

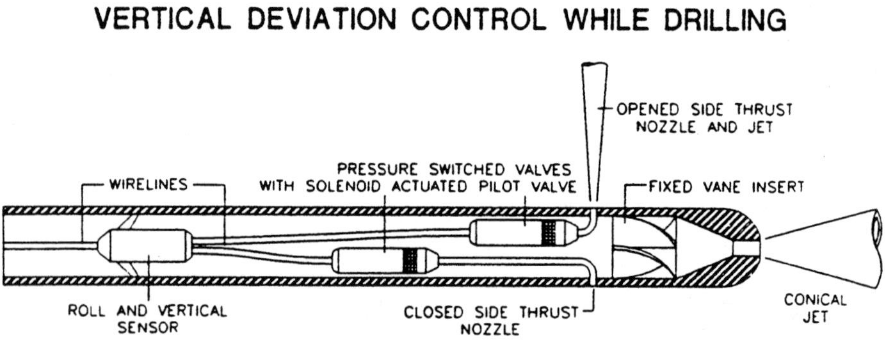

FIGURE 7.11 Directional control obtained by the use of side-thrusting water jets (Dickinson and others, 1992).

Methods Using Coiled Tubing

Recent work has involved the drilling of directional and horizontal holes in which the drilling assembly is attached to a coiled tubing unit rather than to conventional drill pipe. This technique is particularly interesting in reservoirs where there are open fractures that have different reservoir pressures. An example is the Pearsall Field in Texas, in which the producing formation is the Austin Chalk. In this formation, successive vertical fractures may have different fluid pressures, so that if mud density is increased to control high pressure in one fracture, it may result in lost circulation in the next. Conversely, if the mud density is reduced, lost circulation will be reduced, but the well will flow from the high-pressure fractures. Water is the preferred drilling fluid because it does not cause plugging of the fractures at excess pressure and is economical if circulation is lost. It is not, however, dense enough to control pressure by its own hydrostatic head, so drilling is carried out with the wellhead under pressure.

An appropriate diverter system removes the returned fluids, which are then separated. Oil is removed, and the water is returned as the drilling fluid. This is known as "flow drilling." Because of the low bottom-hole pressures, it results in rapid penetration, as well as early revenue from the produced oil.

Under normal circumstances, the well would have to be killed before a trip, which is costly, time consuming, and can cause formation damage.

However, if drilling can be carried out from a coiled tubing unit, the operation can be done under pressure through a snubbing unit without the necessity of killing the well. Fultz and coworkers demonstrated drilling from a coiled tubing unit inside a casing (Fultz and others, 1990) and subsequently described drilling in an open hole (Fultz and Pittard, 1990). This work has recently been followed by the successful drilling of a directional producing well (Ramos and others, 1992). It is expected that these developments will open interesting new possibilities in drilling directional wells.

Directional Drilling in the Near Surface

A large segment of the directional drilling industry is concerned with drilling in the near surface for physical infrastructure purposes, such as the laying of services between buildings and under roads, rivers, or other obstacles. The requirements are characterized by a need for good directional control and for penetration in relatively soft terrain. A subsidiary requirement is that since much of the work is done near inhabited areas, the drilling equipment and methods are preferably nonintrusive. Applications range from the placement of cables from the street to an individual house to the passage of large pipelines under rivers. The latter activity has grown rapidly in recent years; beginning in the late 1970s, about 50 projects totaling 60,000 ft are reported to have been completed by the end of the decade. In 1988, 200 projects, with a total length of 200,000 ft were completed (Hair, 1989a). A world record was recently claimed when a 4,150-ft length of 42-inch gas pipeline was pulled and pushed under the Sacramento River (Anon., 1993a). Another recent job involved pulling 2,000 ft of 48-inch line under the Noord Hollands Canal (Spiekhout and others, 1993). A maximum length of crossing was reported as approximately 6,000 ft of 8-inch line, under the St. Lawrence River near Trois-Rivières (Hair, 1989a).

A rapidly growing application for near-surface directional and horizontal drilling is in the field of environmental investigation and remediation. Special requirements in control, recovery of samples, and cleanliness have to be addressed (BDM International, 1993), and some interesting experiments have already been tried (Kaback and others, 1989). Leo Duffy, former Assistant Secretary of Energy, stated that the use of horizontal wells at the Department of Energy Savannah River Site saved

$125 million, compared to a conventional pump-and-treat scheme for the removal of pollutants (Anon., 1992).

The drilling of large-diameter river crossings is usually carried out by use of a slant rig that is able to push a drill pipe forward at a shallow angle, typically 8 to 12 degrees. The normal method of operation is to drill a pilot hole of a few inches' diameter, using jetting or a conventional directional drilling assembly to change direction. Drilling is normally straightforward—the main consideration being that the operation is frequently carried out in unstable ground composed of unconsolidated sands or clay, although applications in hard rock have been described (McKenny and Knoll, 1989).

In soft ground, drilling may be carried out by using a drill pipe rotating inside a "washover pipe" that is periodically advanced to catch up with the drilling assembly. The washover pipe carries the return mud and cuttings back to the surface so that erosion of the borehole is minimized and the hole is prevented from collapsing. In such a case, the drilling rig is provided with the means to advance and rotate the drill pipe and washover pipes independently. When the hole breaks through on the far side of the crossing, a pulling head with a reaming barrel and fly cutter or "bullet nose" is attached, and this is pulled back through the hole to the starting point, drawing the pipeline behind it. Sometimes, more than one reaming pass may be carried out, resulting in a choice of final pipe pulling direction from either end of the hole (Szczupak, 1989).

Directional control is by motor and bent sub or, in soft ground, by the use of a bit with an eccentric jet nozzle, which can erode one side of the hole preferentially. The steering input may be obtained from a variety of sensors. These may be introduced into the hole (typically on wireline) while drilling is stopped to relay information to the surface ("survey tools"), or they may be made sufficiently robust to be kept in the hole while drilling is in progress. The latter, combined with steering equipment, allow directional changes to be made without interrupting the drilling operations, and are known as "steering tools." For a discussion of these techniques, see Pittard and others (1989).

Different physical principles are employed, including the use of magnetic, induction, and gravity sensing devices, or gyroscopes. Magnetic devices may use the natural gravitational field, or the position of the boring head may be measured by laying out a cable on the surface in a precisely known position above the path of the drill and passing a strong current through it. A magnetometer at the drill head then senses the field generated and provides the necessary steering information. These principles are used in a variety of commercial devices, ranging from real-time measurement while drilling (MWD) to single- and multishot surveying devices to the various tracking methods that detect the position of the drill head from sensors on the surface above the boring head.

Finally, it is important to know the precise position of the "as-built" pipeline. For this application, "gyropigs" are being developed that will record directional data as they are pumped down the pipeline (Spiekhout, 1991). Generally, "in-hole" devices are used in the longer, deeper holes, either because the holes are too deep for the surface sensors to detect the drilling assembly accurately or because the surface locations are inaccessible. Surface detection methods are, however, much less expensive because they do not use complex or delicate "downhole" instruments that have to endure the drilling environment. They are therefore preferred for smaller, shallower holes.

For small-diameter and shorter boreholes, a range of augering and pipe-jacking devices is available. Various nonsteerable augers and "moles" have been developed, using one of three basic excavation methods. Augers excavate the ground and transport the dry cuttings to the rear of the hole by a rotating spiral conveyor. Augers are generally used for holes 8 inches or larger and are useful in loose ground. They can be combined with impact and rotary impact boring heads to deal with harder terrain (Anon., 1990). Moles either operate by wetting the ground to loosen it and then forcing their way forward, or use impact devices to hammer their way into the ground, pushing the soil to the side.

Several mole devices are of particular interest because they are steerable. One such device uses high-pressure jets (25 MPa [megapascals] at 0.03 m^3/s) of a bentonite suspension to drill the hole (Ritchie and others, 1989). This pressure is said to be able to drill most soils but not to damage existing metal structures or concrete. For drilling in harder soils, carbide cutting teeth may be added. The cutter head has a transmitter that sends a signal to a surface locating unit, which is capable of detecting the drill head position at depths down to about 30 ft. The system uses the orientation of the jets in the drilling head to steer the tool.

A family of steerable devices has been developed as the result of a Gas Research Institute project, set up in 1984, that resulted in the development of three small-diameter guided boring systems (Hair, 1989b). The first comprises a pneumatic hammer attached to a rod pushing unit. The penetrating head has an oblique forward surface that causes it to deflect sideways as it penetrates. Behind the head is a sleeve with a pair of spiral fins that engage in the hole wall. The sleeve can be locked in place on the boring head, in which case the penetrating head is forced along a spiral path with no net deviation. If the sleeve is unlocked, however, the head will not rotate, and the hole will be deflected in whatever direction the oblique head is set. For position detection and steering, the drilling head uses a solenoid, emitting a magnetic field that is detected by a surface unit.

The second device is a hydraulically powered rod pusher (Figure 7.12) that has an offset conical penetrating point at the forward end of a string of push rods. There is no drilling motor, and the string is advanced and steered by respectively jacking and rotating it from the jacking pit end (Stangl and Boeckman, 1989).

The third device is again a pipe-jacking system, but with a choice of drilling methods (Pittard and others, 1987; Stangl and others, 1988; Figure 7.13)—either an impact penetrator with oblique face similar to that described above, or a down hole air motor and bent sub developed by Maurer Engineering Co. (Houston) based on motors developed for the petroleum industry.

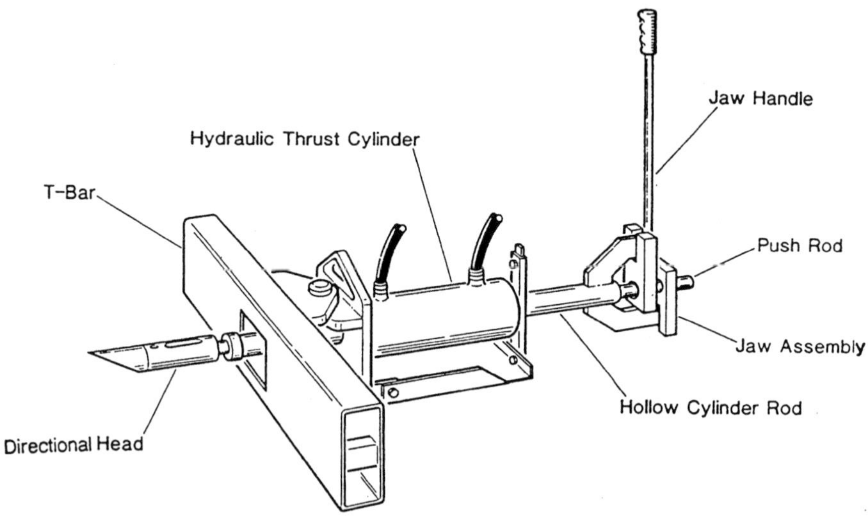

FIGURE 7.12 Rod pushing unit (Stangl and Boeckman, 1989).

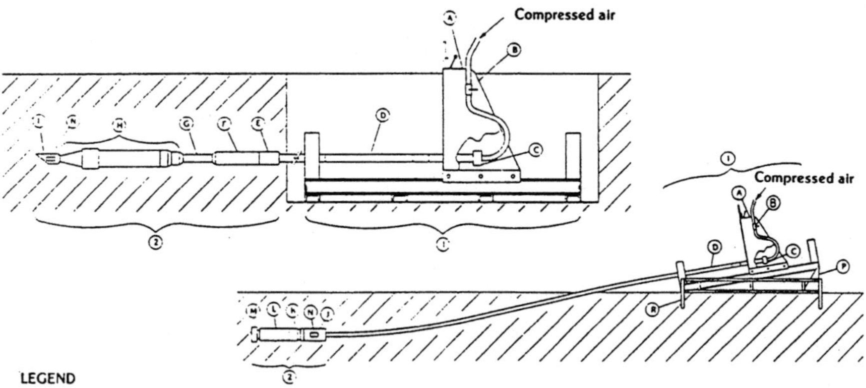

LEGEND

1 Drill frame
2 Percussive downhole tool system
A Carriage assembly
B On-off valve
C Air swivel
D Drill pipe
E Downhole valve
F Shock absorber
G Flex sub
H Percussive mole with connector sub
I Slanted nose steering device and electronics housing
J Bent sub steering device and electronics housing
K Circulating sub
L Air motor
M Drill bit
N Tracking electronics
P Surface launch adaptor
R Anchor posts

FIGURE 7.13 "True Trac" guided drilling equipment (Hair, 1989b).

Priorities for R&D

1. *Directional tunneling*: The great advantage of circular TBMs is that they excavate the whole tunnel face simultaneously and are therefore capable of much greater rates of advance than the selective machines. However, there will always be a demand for tunneling machines that are more flexible in terms of the diameter, and particularly the shape, of tunnels that are cut. In most cases, it is desirable to have a flat floor in order to allow access by wheeled or tracked transport (water and some rail tunnels being an exception). Designs that allow the driving of noncircular tunnels, but simultaneously permit excavation to proceed over most of the tunnel face, are of great utility. These will allow high power levels to be brought to the working face, with attendant high rates of advance. If mechanical excavation methods are used, a heavy and bulky drive system will be needed, which makes sharp turns difficult. This is the principal reason that TBMs are frequently used to drive transportation or water

tunnels, which are usually only gently curved but are seldom found in mines. Research should focus on designs that are more compact, to allow easier steering while at the same time allowing simultaneous excavation of the whole face. In the longer term, if a nonmechanical means for excavating the tunnel can be developed, it may not be necessary to have a large, rigid supporting structure, nor will it be necessary to maintain the circular cross section. Continued long-term research that focuses on novel excavation methods for directional tunneling should also be pursued.

2. *Directional drilling*: The subject of directional drilling covers many fields, each having its own particular demands. In general, there will always be a need for adaptable methods that allow holes of different diameters and degrees of curvature to be drilled with the same equipment. The wide range of methods available in the petroleum and civil engineering sectors, each with its own particular niche application, illustrates the need for the development of more general directional drilling methods that will have a wider range of use. Specific research needs for directional drilling include the following:

- *Directional drilling and sharp curves*: Methods are needed to drill sharper curves in holes with larger diameters. Since one of the limiting factors in this case is the difficulty of making large-diameter tubulars (particularly casings) curve around tight turns, research should be undertaken to develop flexible casings and liners, or to develop methods for forming a curved casing in situ. For similar reasons, research is needed to develop drill strings that will transmit high loads around the curve and along the horizontal portion of the well, but will be resistant to buckling and to flexural fatigue. Lightweight alloys or novel composite materials offer interesting possibilities in this area.
- *Directional drilling and lubricating muds*: Transmission of drilling forces to the hole bottom in near-horizontal wells will be aided by the development of smooth drilling assemblies, helped by highly lubricating muds. Research in this area should focus on muds that have good cuttings-carrying capacity. In the particular case of the growing market for environmental drilling, the muds should have low toxicity.
- *Drill bits*: Since transmission of weight to the drill bit is critical in much directional drilling, particularly extended reach drilling, there is a need for research to develop drill bits that have high cutting ability at low weight on the bit. Such very aggressive bits may have to be

protected from the application of excessive weight if this becomes available, pointing to the need for a downhole system of weight and/or torque control.

- *Bottom-hole assemblies*: An alternative approach to drill bits might be to develop a BHA that grips the hole wall and pulls itself along. This would eliminate the need for many of the specialized arrangements of collars uphole, compressive service drill pipe, and limitations to hole curvature that arise from having to transmit compressive forces around the curve. The committee recognizes that some work was undertaken in this area a few years ago and recommends continued research and development on BHA alternatives. For example, the use of coiled tubing for directional drilling deserves emphasis for continued development.
- *Directional sensing instrumentation*: Research is needed to develop more accurate and cheaper direction-sensing instrumentation, particularly for equipment that can detect and steer the drilling equipment without the need to stop drilling operations. Thus, locating equipment mounted on the surface will be preferred over in-hole instruments. Current technology does not, however, allow accurate measurements to be made at great depths. Until such methods become available, downhole measurements will have to be maintained. For these, continuous measuring and transmitting systems (MWD) will be preferred (i.e., steering tools rather than survey tools), particularly if these are combined with the possibility of making a wide range of steering corrections.
- *Smart directional drilling*: The committee endorses research efforts that combine geologic inputs with directional information to allow smart drilling, for example, to follow a particular oil-bearing formation at depth, or to seek or avoid stable or unstable, polluted or clean strata for directional services or environmental projects. Research should be undertaken to develop sensor/guidance systems that avoid service lines or other buried structures in the near surface. Ultimately, it may be possible to build "closed-loop" systems that take the appropriate avoiding actions and then return to the predetermined track without operator intervention.
- *Inclined borehole stability*: Highly inclined boreholes are inherently less stable than those that are near vertical. This has long been recognized in the petroleum sector and is no less a problem in near-surface drilling. Research should focus on the stabilization of inclined boreholes, particularly in the near surface. The environmental drilling sector will have

particular requirements for low-toxicity, minimally invasive methods. Ground freezing while drilling may offer one solution.

- *Directional drilling in poor ground*: Near-surface applications must deal with ground that not only is poorly cemented but contains a mix of strong and weak components, such as sand or clay with boulders. Not only does this type of ground give great problems with borehole stability, it also makes maintaining borehole direction extremely difficult, and tool wear in such mixed ground is notorious. Research should be undertaken to study and develop ways to drill satisfactorily in poor ground material.

References

Akiba, T., and Yamada, T., 1989, The Enviliner Microtunnelling system: NO-Dig 89, Fourth International Conference and Exhibition on Trenchless Construction for Utilities, London, ISTT, p. 243-245.

Anon., 1988, Underground microtunnel index: Underground, March, p. 28-31.

Anon., 1989, Horizontal systems growing more specialized, parts I and II: Offshore, v. 49 (Oct./Nov.), p. 36-43.

Anon., 1990, Trenchless technology today: Pipeline and Gas Journal, v. 216 (Feb.), p. 33-36.

Anon., 1992, Nuclear Waste News, v. 12 (7), p. 57.

Anon., 1993a, Pipeline river crossing record set in California: Oil and Gas Journal, v. 91 (5), p. 56.

Anon., 1993b, Statoil claims world record in extended reach: Oil and Gas Journal, v. 91 (7), p. 31.

Anon., 1993c, Texaco claims record for horizontal displacement: Oil and Gas Journal, v. 91 (9); p. 34.

Austin, C. E., Rose, R. E., and Schuh, F. J., 1988, Simultaneous multiple entry hydraulic fracture treatments of horizontally drilled wells: SPE 18263, SPE Annual Technical Conference and Exhibition, Houston, Tex., Oct. 2-5, Richardson, Tex., SPE, p. 811-826.

Barrett, S. L., and Lyon, R. G., 1988, The Navigation Drilling System proves effective in drilling horizontal wells in the Java Sea: SPE 17238, IADC/SPE Drilling Conference, Dallas, Tex., Feb. 28-Mar. 2, Richardson, Tex., SPE, p. 503-510.

BDM International, 1993, Evaluation of Alternative Technologies and Subsurface Confinement Barriers for Single Shell Tanks at Hanford, Morgantown, West Virginia, Report prepared for the U.S. Department of Energy, 226 pp.

Beaumont, T. C., 1989, UK experiences and developments of the Witte microtunelling machines: NO-Dig 89, Fourth International Conference and Exhibition on Trenchless Construction for Utilities, London, ISTT, p. 237-242.

Bosio, J. C., Fincher, R. W., Giannesini, J. F., and Hatten, J. L., 1987, Horizontal drilling—a new production method: 12th World Petroleum Congress, Houston, Tex., April, Chinchester, John Wiley and Sons, v. 3, p. 187-199.

Cooper, G. A., 1994, Directional drilling: Scientific American, v. 270 (5), p. 82-87.

Dawson, R., and Paslay, P. R., 1984, Drillpipe buckling in inclined holes: Journal of Petroleum Technology, v. 36 (10), p. 1734-1738.

Dickinson, W., and Dickinson, R. W., 1985, Horizontal radial drilling system: SPE paper 13949, SPE annual California Regional Meeting, Bakersfield, Calif., May 27-29, Richardson, Tex., SPE, p. 887-892.

Dickinson, W., Anderson, R. R., Dickinson, R. W., and Dykstra, H., 1987, Gravel packing of horizontal wells: SPE 16931, SPE Annual Technical Conference and Exhibition, Dallas, Tex., Sep. 27-30, Richardson, Tex., SPE, p. 519-528.

Dickinson, W., Anderson, R. R., and Dickinson, R. W., 1989, The ultrashort-radius radial system: SPE Drilling Engineering, v. 4 (3), p. 247-254.

Dickinson, W., Dykstra, H., Nees, J. M., Dickinson, E., 1992, The ultrashort radius radial system applied to thermal recovery of heavy oil: SPE Western Regional Meeting, Bakersfield, Calif., March 30-April 1, Richardson, Tex., SPE, p. 583-600.

Eastman, H. J., 1954, Lateral drainhole drilling: The Petroleum Engineer, v. 26 (12), p. B57-B73.

Eaton, N., 1990, Newly developed systems meet demands of coring: World Oil, v. 210 (3), p. 62-66.

Eddison, A., and Symons, J., 1990, Downhole adjustable gauge stabilizer improves drilling efficiency in directional wells: SPE 20454, 65th Annual Technical Conference and Exhibition, New Orleans, La., September 23-26, Richarson, Tex., SPE, p. 509-516.

Edlund, P. A., 1987, Application of recently developed medium curvature horizontal drilling technology in the Spraberry Trend area: SPE 6170, SPE/IADC Drilling Conference, New Orleans, La., March 15-18, Richardson, Tex., SPE, p. 1057-1063.

Fultz, J. D., and Pittard, F. J., 1990, Open hole drilling using coiled tubing and a positive displacement mud motor: SPE 20459, 65th Annual Technical Conference and Exhibition, New Orleans, La., Sep. 23-26, Richardson, Tex., SPE, p. 551-559.

Fultz, J. D., Pittard, F. J., Sawyer, F. D., and Farmer, F. D., 1990, Slim-hole drilling in harsh environments: IADC/SPE 19949, IADC/SPE Drilling Conference, Houston, Tex., Feb. 27-Mar. 2, Richardson, Tex., SPE, p. 333-340.

Gibson, J., 1993, Robbins Co, Kent Wash., Personal communication.

Glagola, M. A., and Wong, L. F., 1986, Aluminum drillpipe for directional drilling: SPE/IADC 14789, IADC/SPE Drilling Conference, Dallas, Tex., Feb. 10-12, Richardson, Tex., SPE, p. 553-560.

Haas, R. C., and Stokley, C. O., 1989, Drilling and completing a well in fractured carbonate, World Oil, v. 204 (4), p. 39-45.

Hair, J. D., 1989a, River crossing technology: Pipeline and Gas Journal, v. 215 (1), p. 29-35.

Hair, J. D., 1989b, Small diameter guided boring systems: Pipeline and Gas Journal, v. 215 (4), p. 18-24.

Hayashi, M., and Miyata, Y., 1989, Okumura—Markham super mini shield tunneling method: NO-Dig 89, Fourth International Conference and Exhibition on Trenchless Construction for Utilities, London, ISTT, p. 231-235.

Hourcard, M., and Bannerman, J., 1990, Beware of sloughing shales overlying horizontal holes: World Oil, v. 211 (1), p. 101-105.

Joshi, S. D., 1987a, A review of horizontal well and drainhole technology: SPE 16868, SPE Annual Technical Conference and Exhibition, Dallas, Tex., Sep. 27-30, Richardson, Tex., SPE, p. 339-354.

Joshi, S. D., 1987b, A review of thermal oil recovery using horizontal wells: In Situ, v. 11 (2/3), p. 211-259.

Joshi, S. D., 1988, A review of horizontal well and drainhole technology: SPE 16868, Rocky Mountain Regional Meeting, Casper, Wyo., May 11-13, Richardson, Tex., SPE, p. 89-108.

Kaback, D. S., Looney, B. B., Corey, J. C., Wright, L. M., and Steele, J. L., 1989, Horizontal wells for in situ remediation of groundwater and soils: National Water Well Association Outdoor Action Conference, Orlando, Fla., May 22-25, 16 pp.

Karlsson, H., Cobbley, R., and Jaques, G. E., 1989, New developments in short-, medium-, and long-radius lateral drilling: SPE 18706, SPE/IADC Drilling conference, New Orleans, La., Feb. 28-Mar. 3, Richardson, Tex., SPE, p. 725-736.

Keelean, L., Harris, S. S., and Petronio, N., 1989, Short radius drilling technology utilizing mobile service/workover rig: SPE/IADC 18711, SPE/IADC Drilling Conference, New Orleans, La., Feb. 28-Mar. 3, Richardson, Tex., SPE, p. 765-772.

Lattimore, G. M., Carden, R. S., and Fischer, T., 1987, Grand Canyon directional drilling and waterline project: SPE/IADC 16169, SPE/IADC Drilling Conference, New Orleans, La., Mar. 15-18, Richardson, Tex., SPE, p. 1047-1056.

Logan, T. L., 1988, Horizontal drainhole drilling techniques used for coal seam resource exploitation, SPE 18524, SPE 63rd Annual Technical Conference and Exhibition, Houston, Tex., 2-5 October, Richardson, Tex., SPE, p. 711-723.

Logan, T. L., Schwoebel, J. J., and Horner, D. M., 1987, Application of horizontal drainhole drilling technology for coalbed methane recovery: SPE/DOE 16409, SPE/DOE Low Permeability Reservoirs Symposium, Denver, Colo., May 18-19, Richardson, Tex., SPE, p. 195-206.

Lubinski, A., 1950, A study of the buckling of rotary drilling strings: Drilling and Production Practice, API, p. 178.

Makohl, F., and Jurgens, R., 1986, Evolution and differences of directional and high-performance downhole motors: IADC/SPE Drilling Conference, Dallas, Tex., Feb. 9-12, Richardson, Tex., SPE, p. 189-196.

McKenny, F. S., and Knoll, W. G., 1989, Hard rock directional crossings problems solved with new directional drilling systems: NO-Dig 89, Fourth International Conference and Exhibition on Trenchless Construction for Utilities, London, ISTT, p. 211-216.

Millheim, K. K., and Warren, T. M., 1978, Side cutting characteristics of rock bits and stabilizers while drilling: SPE 7518, SPE Annual Technical Conference and Exhibition, Houston, Tex., Oct. 1-3, Richardson, Tex., SPE, v. 7 (4).

Millheim, K. K., and Apostal, M. C., 1981, The effect of bottomhole assembly dynamics on the trajectory of a bit: Journal of Petroleum Technology, v. 33 (12), p. 2323-2338.

Miyazaki, S., 1993, Personal communication.

Moria, H., and Sawaguchi, A., 1989, Microtunelling system with directional control and long distance driving: NO-Dig 89, Fourth International Conference and Exhibition on Trenchless Construction for Utilities, London, ISTT, p. 273-282.

Parsons, R. S., and Fincher, R. W., 1986, Short-radius lateral drilling: a completion alternative: SPE 15943, SPE Eastern Regional Meeting, Columbus, Ohio, Nov. 12-14, Richardson, Tex., SPE, p. 239-250.

Pittard, G. T., Macdonald, W. J., Kramer, S. R., Wasson, M. R., and Herben, W. C., 1987, Development and application of guided horizontal boring equipment for utility applications: NO-Dig 87, Second International Conference and Exhibition on Trenchless Construction for Utilities, London, ISTT, p. 2.5.1-2.5.8.

Pittard, G. T., McDonald, W. J., and Kramer, S. R., 1989, Instrumentation systems for guided boring: NO-Dig 89, Fourth International Conference and Exhibition on Trenchless Construction for Utilities, London, ISTT, p. 191-199.

Ramos, A. B., Fahel, R. A., Chaffin, M., and Pulis, K. H., 1992, Horizontal slim-hole drilling with coiled tubing: an operator's experience: IADC/SPE 23875, SPE/IADC Drilling Conference, New Orleans, La., Feb. 18-21, Richardson, Tex., SPE, p. 289-300.

Rehm, W. A., and Garcia, A., 1989, Horizontal drilling in mature oil fields: SPE/IADC 18709, SPE/IADC Drilling Conference, New Orleans, La., Feb. 28-Mar. 3, Richardson, Tex., SPE, p. 755-764.

Reiss, L. H., 1987, Production from horizontal wells after five years: Journal of Petroleum Technology, v. 39 (11), p. 1411-1416.

Reiss, L. H., Jourdan, A. P. L., Giger, F. M., and Armessen, P. A., 1984, Offshore and onshore European horizontal wells: Paper OTC 4791, Offshore Technology Conference, Houston, Tex., May 7-9, p. 137-144.

Ritchie, J., Enenbach, J., Kleiser, K., and Steutel, M. J., 1989, Use of the FlowMole GuideDril system in Europe: NO-Dig 89, Fourth International Conference and Exhibition on Trenchless Construction for Utilities, London, ISST, p. 283-289.

Robbins Co., 1993, Operating instructions for model 167-266/267 TBM: Robbins Co. Kent, Wash., p. 2.21-2.40.

Schuh, F. J., 1989, Horizontal well planning—build curve design: SPE 20150, Petroleum Technology into the Second Century, New Mexico

Tech, Socorro, N. Mex., Oct. 16-19, Richardson, Tex., SPE, p. 47-61.

Shale, L., 1991, Development of air drilling motor holds promise for specialized directional drilling applications: SPE 22564, SPE 66th Annual Conference and Exhibition, Dallas, Tex., Oct. 6-9, Richardson, Tex., SPE, p. 275-286.

Sheikholeslami, B. A., Schlottman, B. W., Siedel, F. A., and Button, D. M., 1989, Drilling and production aspects of horizontal wells in the Austin chalk: SPE 19825, SPE Annual Technical Conference and Exhibition, San Antonio, Tex., Oct. 8-11, Richardson, Tex., SPE, p. 575-590.

Spiekhout, J., 1991, Developments in directional drilling: Pipeline and Gas Journal, v. 215 (4), p. 54-57.

Spiekhout, J., Gorter, J., and Hartlief, W. H., 1993, Records set, drilling calculation refined in Dutch pipeline crossings: Oil and Gas Journal, v. 91 (1), p. 33-38.

Stacey, E. S., Rogers, C. T., Bowman, J. K., Noblett, B. R., and Cooney, M. F., 1992, Case history of a horizontal well using a high angle pilot hole: IADC/SPE 23880, IADC/SPE Drilling Conference, New Orleans. La., Mar. 18-21, Richardson, Tex., SPE, p. 331-340.

Stangl, G. A., Lee, D. W., and Wassen, M. R., 1988, Extended range horizontal boring using the TrueTrac directional boring system: NO-Dig 88, Third International Conference and Exhibition - Achievements in Trenchless Methods, Washington, D.C., ISTT, p. 319.

Stangl, G. A., and Boeckman, W. G., 1989, Field experiences with directional rod pushers: NO-Dig 89, Fourth International Conference and Exhibition on Trenchless Construction for Utilities, London, ISTT, p. 261-272.

Stormont, D. H., 1953, Increased drainage of oil into well by drainhole drilling: Oil and Gas Journal, v. 52, (11), p. 105-108.

Szczupak, J. R., 1989, Horizontal directional drilling applications: NO-Dig 89, Fourth International Conference and Exhibition on Trenchless Construction for Utilities, London, ISTT, p. 201-210.

Tolle, G., and Dellinger, T. B., 1986, Mobil identifies extended reach drilling advantages, possibilities in North Sea: Oil and Gas Journal, v. 84 (21), p. 78-86.

White, C. W., 1989, Drilling and completion of a horizontal lower Spraberry well including multiple hydraulic fracture treatments: SPE

19721, 64th Annual Technical Conference and Exhibition, San Antonio, Tex., Oct. 8-11, Richardson, Tex., SPE, p. 205-210.

Whiteley, M. C., and England, W. P., 1985, Air drilling operations improved by percussion bit/hammer tool tandem: SPE/IADC 13429, SPE/IADC Drilling Conference, New Orleans, La., Mar. 6-8, Dallas, Richardson, Tex., SPE, p. 61-70.

Wilson, R. C., and Willis, D. N., 1986, Successful high angle drilling in the Statfjord field: SPE 15465, 61st SPE Annual Technical Conference and Exhibition, New Orleans, La., Oct. 5-8, Richardson, Tex., SPE, 14 pp.

Yost, A. B., Overbey, W. K., and Carden, R. S., 1987, Drilling a 2000-ft. horizontal well in the Devonian shale: SPE 16681, SPE Annual Technical Conference and Exhibition, Dallas, Tex., Sep. 27-30, Richardson, Tex., SPE, p. 291-302.

8
BOREHOLE STABILITY

Introduction

Borehole stability technology includes chemical as well as mechanical methods to maintain a stable borehole, both during and after drilling. Drilling fluids range from water to oil to complex chemical systems with properties designed for specific site conditions to aid the drilling process. Drilling fluids perform the following functions: carrying cuttings out of the hole, cleaning the bit, cooling and lubricating the bit, providing buoyancy to the drill string, controlling formation fluid pressures, preventing formation damage, and providing borehole support and chemical stabilization.

Most borehole stability and drilling fluids-related problems can be handled with present technology in relatively easy, well-defined environments, provided stringent quality control measures are maintained. Nevertheless, severe, complex drilling situations and formations still present serious challenges to economically viable drilling.

Borehole failures are an increasing concern due to an "explosion" in drilling horizontal wells in unique geological environments (e.g., Hanford site) as well as drilling "difficult" hydrocarbon reservoirs. Difficult reservoirs include, for example, unconsolidated or poorly consolidated sediments, shales, complex reservoir geometries, naturally fractured reservoirs, and overpressured reservoirs. This had led to greater research emphasis related to the stability of circular rock openings (International Society for Rock Mechanics, 1987; National Research Council, 1993). Borehole stability has been the subject of several recent research papers (Guenot, 1987; Maury, 1987; Maury and Sauzay, 1987; Santarelli and Brown, 1987; Bjarnasson and others, 1988; Roegiers and Detournay, 1988; Beus and Dar, 1989; Ewy and Cook, 1989; Haimson and Herrick, 1989; Périé and Goodman, 1989), and current borehole design criteria have been reassessed (Detournay and Fairhurst, 1987; Gill and Ladanyi, 1987; Gumusoglu and others, 1987; Mühlhaus, 1987; Lin and Fairhurst, 1988; Vardoulakis and Papanastasiou, 1988; Papanastasiou and Vardoulakis, 1989).

The goal of this research is to provide an efficient basis to obtain a near-gauge opening that can be maintained easily for the long term with minimal environmental impact, disturbance of the drilling objectives, and drilling costs. In some industries (e.g., petroleum and environmental), the emphasis on the near-gauge opening is to minimize the costs of drilling, evaluating, and cementing the hole, whereas for other applications this could very well be seen as an attempt to mobilize the residual rock strength in order for it to participate in the overall support requirements. Long term implies no "unexpected" maintenance costs for the life of a project.

Efforts and progress are being made by several organizations and have led to new, important, proprietary, or patented technologies usually available for license, but rarely utilized by large numbers of laboratories or applied in the field. For example, one oil company has recently developed an extensive shale data base along with a patented method to determine in situ shale strengths from correlations with index properties derived from simple measurements on rock cuttings. Included in this data base are correlations to assist in providing proper mud weight guidelines for several high-angle oil and gas well projects.

When considering areas in which improvements could be made to avoid stability or environmental problems associated with the use of drilling fluids, the following are selected research needs (in no particular priority order):

- *Rock-fluid interaction*: A better understanding of this coupled phenomenon may lead to greater penetration rates.
- *Flow balance measurements:* Unexpected loss of drilling fluids can lead to catastrophic failures; hence, any advanced warning, especially in geothermal environments, will be beneficial. Also, the development of drilling fluids that result in zero fluid loss, no matter what the formation characteristics, would be breakthrough.
- *Air-based systems*: Such systems could decrease formation damage and address some of the environmental concerns, provided dust can be adequately controlled.

- *Heterogeneous media*: Substantial changes in formation characteristics affect drilling orientation and efficiency. Sampling in such environments is also challenging.
- *Hostile environments*: Permafrost and hot formations, for example, adversely affect the penetration rate.
- *Environmentally acceptable drilling fluids:* Nontoxic, oil-like fluids to avoid undue formation damage are needed.
- *Formation characterization*: Characterization methods are needed for environmental applications, especially contaminant detection and measurement. Characterization methods are also needed for shales because a majority of hydrocarbon well footage is drilled in shaley lithologies. This includes both pertinent rock properties and reservoir characteristics, such as the existing stress field and pore pressure.
- *Reliable stability models*: Stability models, especially models that couple thermal, porous, mechanical, and chemical phenomena, are needed.

Priorities for R&D

From the above list, six important research areas could yield significant advances and benefits. As with most engineering disciplines, there exists a wide gap between R&D developments and field applications. In addition to new research, efforts should be made to transfer some of the existing technologies that could immediately improve problems encountered in borehole stability.

1. *Formation characterization and validation*: Formation characterization should be undertaken during the drilling process, integrating the information obtained from drilling data, various types of logs, cutting data, seismic information, surface measurements, structural geology, and in situ stress measurements. The critical parameters that should be determined are the following:

- rock strength, usually at in situ conditions of pressure and temperature; an attempt could be made to determine its variation during the drilling process due to progressive unloading of the rock;

• formation constitution, because some minerals such as smectite could cause time-dependent instabilities due to environmental changes;

• permeability, because it governs potential inflow (production) and outflow (loss of drilling mud) conditions (it should be realized that for heavily fractured reservoirs, in which secondary permeability is greatly affected by the applied stresses, care should be taken when assessing fluid transmissivity from borehole geometries);

• pore pressure, because it affects safety (formation fluid and gas kicks) and influences all hydraulic and formation constitutive effects;

• stress state, which controls most mechanisms;

• abrasivity, which influences drillability; and

• discontinuities, because they introduce major heterogeneities in all aspects.

2. *Rock alteration*: A very limited amount of research has been undertaken on the chemomechanical weakening of rocks. The underlying mechanisms are still very much disputed; therefore, the potential for defining and controlling this rock-fluid interaction has not been fully realized. The same is true for drilling in very sensitive formations for which specific inhibitive fluids could be developed. Finally, injection of "active" fluids at the rock face to increase penetration rate and decrease bit wear has not been pursued aggressively. One might envision two flow loops, a primary one with the purpose of transporting the cuttings and a secondary one to help the drilling process.

3. *Improved stability models and rock data (shale)*: Existing design codes for stability models are quite conservative. For example, linear elasticity would predict that some vertical and many directional boreholes drilled for hydrocarbon exploration and production should be unstable when, in fact, many are stable. The present limitation is due not only to the models used, but also to the poor knowledge of the constitutive behavior of the formation. In addition, time-dependent failures that occur occasionally (e.g., during the production phase of a reservoir) are almost never predicted, pointing to the need for introducing solutions that take into account the coupling between fluid flow and rock deformation. Such poroelastic efforts are currently being pursued. In all cases, the design codes need to be user friendly and easy to apply.

4. *Drilling techniques for environmental remediation boreholes*: Drilling techniques for environmental remediation are evolving as emphasis is placed on removing subsurface contamination from Superfund and other contaminated sites. Current approaches are somewhat limited, but the industry is borrowing some existing petroleum technology. The development of new technology or the adaptation of existing technologies from other industries is required to significantly expand current economics. Expanded and improved technology is necessary for horizontal wells; reverse circulation; dry-air drilling; and drilling waste/effluent minimization, recovery, and disposal.

5. *Methods for drilling and stabilizing heterogeneous formations*: Heterogeneous formations present special and expensive difficulties both in drilling and in maintaining a stable borehole. Formations that contain both hard (or abrasive) and soft components drill very slowly with severe demands on bits. Fractured or heavily jointed rock causes more complex loading of the borehole that requires, in many instances, the development of more sophisticated numerical modeling to predict borehole behavior.

6. *Drilling fluids*: Drilling fluids technology is a moving target due to rapidly expanding needs, demands, and restrictions such as environmental remediation, air drilling, severe temperature conditions, increased lubricity requirements, restrictions on oil-base systems, discharge limitations, and horizontal and extended reach drilling.

For example, oil-base drilling fluids have come under severe regulatory restrictions. Effective alternatives have been developed, but they are expensive. To comply with new government regulations restricting the use of some technologies or practices, drilling organizations have responded well by developing acceptable alternatives; however, these solutions usually have substantial added costs and limitations that are sometimes prohibitive.

Fluid development needs encompass the design of new environmentally acceptable water-base fluids and oil-like systems that will provide alternatives to oil-base devices. Such new fluids should provide superior filtration control to minimize fluid invasion and damage to permeable zones. They should possess good mudcake properties to provide needed filtration control and prevent differential-pressure sticking of the drill pipe against the borehole wall. The new fluids should provide adequate hole-cleaning capabilities in horizontal or high-angle wells. Improvements in understanding the fundamentals of cutting transport flow visualization, air-foam behavior, and fluid viscoelastic behavior will aid that process.

References

Beus, M. J., and Dar, S. M., 1989, Three-dimensional FEM analysis to scale field measurements from deep mine accessways: *in* Rock Mechanics as a Guide for Efficient Utilization of Natural Resources, Proceedings 30th U.S. Symposium on Rock Mechanics, Morgantown, W. Va., p. 783-790.

Bjarnasson, B., Ljunggren, C., and Stephansson, O., 1988, New developments in hydrofracturing stress measurements at Lulea University of Technology: Proceedings 2nd Workshop on Hydraulic Fracturing Stress, Measurements (HFSM'88), University of Minnesota, v. 1, p. 113-140.

Detournay, E., and Fairhurst, C., 1987, Two-dimensional elastoplastic analysis of a long cylindrical cavity under non-hydrostatic loading: International Journal of Rock Mechanics, Mineral Science, and Geomechanical Abstracts, v. 24 (4), p. 197-211.

Ewy, R. T., and Cook, N. G. W., 1989, Fracture processes around highly stressed boreholes: Proceedings 12th Annual Energy-Sources Technical Conference, Houston, ASME, New York, p. 63-70.

Gill, D. E., and Ladanyi, B., 1987, Time-dependent ground response curves for tunnel lining design: Proceedings 6th ISRM Congress, Montreal, Rotterdam, Balkema, p. 917-921.

Guenot, A., 1987, Contraintes and ruptures autour des forages petroliers: Proceedings 6th ISRM Congress, Montreal, Rotterdam, Balkema, p. 109-118.

Gumusoglu, M. C., Bray, J. W., and Watson, J. O., 1987, Analysis of underground excavations incorporating the stain softening of rock masses: Proceedings 6th ISRM Congress, Montreal, Rotterdam, Balkema, p. 923-928.

Haimson, B. C., and Herrick, C. G., 1989, Borehole breakouts and in-situ stress: Proceedings 12th Annual Energy-Sources Technical Conference, Houston, ASME, New York, p. 17-22.

International Society for Rock Mechanics, 1987, Failure mechanisms around underground excavations: Commission Report, Proceedings

6th ISRM Congress, Montreal, Rotterdam, Balkema, p. 1119-1128.

Lin, D., and Fairhurst, C., 1988, Static analysis of the stability of three-dimensional block systems around excavations in rock: International Journal of Rock Mechanics, Mineral Science, and Geomechanical Abstracts, v. 25, p. 139-147.

Maury, V., 1987, Observations, recherches et resultats recents sur les mecanismes de ruptures autour de galeries, isolees—Report of the ISRM Commission on Failure Mechanisms Around Underground Excavations: Proceedings 6th ISRM Congress, Montreal, Rotterdam, Balkema, p. 1119-1128.

Maury, V., and Sauzay, J. M., 1987, Borehole stability: case histories, rock mechanics approach and results: Proceedings SPE/IADC Conference, New Orleans, SPE 16051, Richardson, Tex., SPE, p. 11-24.

Mühlhaus, H. B., 1987, Stability of deep underground excavations in strongly cohesive rock: Proceedings 6th ISRM Congress, Montreal, Rotterdam, Balkema, v. 2, p. 1157-1161.

National Research Council, 1993, Stability, Failure, and Measurements of Boreholes and Other Circular Openings: Washington, D.C., National Academy Press, 92 pp.

Papanastasiou, P., and Vardoulakis, I., 1989, Bifurcation analysis of deep boreholes—II. Scale effect: International Journal of Numerical Analysis and Methods of Geomechanics, v. 13, p. 183-198.

Périé, P. J., and Goodman, R. E., 1989, Evidence of new failure patterns in a thick-walled cylinder experiment: Proceedings 12th Annual Energy-Sources Technical Conference, Houston, New York, AMSE, p. 23-27.

Roegiers, J. C., and Detournay, E., 1988, Consideration on failure initiation in inclined boreholes: Proceedings 29th U.S. Symposium on Rock Mechanics, University of Minnesota, p. 461-469.

Santarelli, F. J., and Brown, E. T., 1987, Performance of deep wellbores in rock with a confining pressure-dependent elastic modulus: Proceedings 6th ISRM Congress, Montreal, Rotterdam, Balkema, p. 1217-1222.

Vardoulakis, I., and Papanastasiou, P., 1988, Bifurcation analysis of deep boreholes—I. Surface instabilities: International Journal of the Numerical Analysis and Methods of Geomechanics, v. 12, p. 379-399.

9
MATERIALS HANDLING

Introduction

Materials handling involves the relocation of materials when a hole is drilled or excavated. The current practice of materials handling depends on the type of hole being constructed (e.g., wells, mine shafts, or tunnels); the site conditions; and the size, orientation, and length of the hole. In general, materials must be transported from the bit face to the surface and from the surface to disposal. Materials handling limits the rate of hole advance when materials cannot be transported to the surface as rapidly as they are mined or when they cannot be moved from the surface to a disposal area as rapidly as they are brought to the surface. Surface disposal problems often relate to environmental problems associated with toxic materials (liquids or solids). The problems that restrict the movement of materials from the bit face to the surface are often the result of material movement interfering with other functions that occur in the same space. For example, in a tunneling operation, the handling of large amounts of materials might interfere with other uses of the tunnel such as transporting personnel and supplies. In drilling, the materials may not fit in the annular space between the hole and the drill pipe.

Status of the Field for Drilling Wells

Rotary drilling, the most common process for drilling wells, uses circulating fluids (both liquids and air) to remove the drilled solid cuttings. Other removal methods include screw augers, which continuously remove solids; buckets; and bailers, which allow water and solids to fill a tube that is periodically retrieved and emptied. These methods are seldom used and are not be described here. In the circulating fluid method (see Figure 2.1), the fluid is picked up by the pump, circulated through the surface piping, and sent down the inside the drill pipe and drill collars, where it exits the bit and entrains the drilled cuttings. The fluid carries the cuttings to the surface for separation and disposal. The behavior of the cuttings in the circulating fluid depends on fluid rheology, as shown in Figure 9.1. The different behaviors of the fluid help determine the degree of difficulty in removing the cuttings from the hole (Walker and Mayes, 1975).

The ability to use fluids to transmit solids depends on fluid properties such as density, viscosity, and velocity (Bourgoyne and others, 1986). Because cuttings are usually more dense than fluids, they will fall under the influence of gravity. As the density difference between fluids and solids decreases, the rate of fall slows. Consequently, the higher the fluid density, the easier it is to remove the solids from the hole. Similarly, as the viscosity of the fluid increases, the difference in velocity between the fluid and solids decreases. Consequently, the higher the fluid viscosity, the easier it is to remove solids from the hole. The higher the fluid velocity, the more rapidly solids are removed from the hole. The velocity must, at a minimum, exceed the rate of fall of the solids for the solids to have a positive velocity out of the hole.

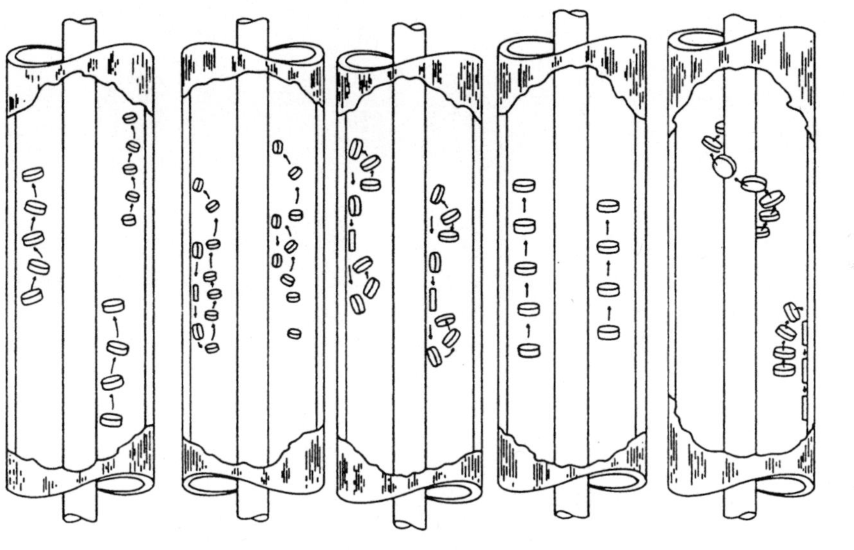

Part A
Discs in turbulent flow
(Stationary center pipe)

Part B
Small discs in laminar flow
(Stationary center pipe)

Part C
Medium discs in laminar flow
(Stationary center pipe)

Part D
Large discs in laminar flow
(Stationary center pipe)

Part E
Medium discs in laminar flow
(Rotating center pipe)

FIGURE 9.1 Levitation of solids in liquid solutions (Moore, 1974).

Unfortunately, fluid properties such as density, viscosity, and velocity serve other purposes that limit the range of properties that can be used. For instance, if the density is too high, circulation will be lost; if it is too low, the well will flow. Higher viscosities reduce the drilling rate and can also cause greater friction-induced pressure buildups, which in turn lead to loss of circulation. Excess velocity can erode the hole wall and produce pressure drops that may lead to loss of circulation.

Circulating fluids serve purposes other than removal of drilled cuttings, namely, density for well control, chemical stabilization of the rock, prevention of filtrate invasion of the rock, cooling and lubricating the bit, suspension of cuttings and weight materials, and corrosion prevention (Rodgers, 1963). These functions must be maintained as the removal of drilled solids is accomplished. Interference among these functions is one of the major problems faced when using a circulating fluid to remove drilled cuttings.

Once the cuttings have been circulated to the surface, they must be separated from the fluid. Figure 9.2 shows an example of a system used to remove drilled cuttings from the fluid. The drilling fluid containing

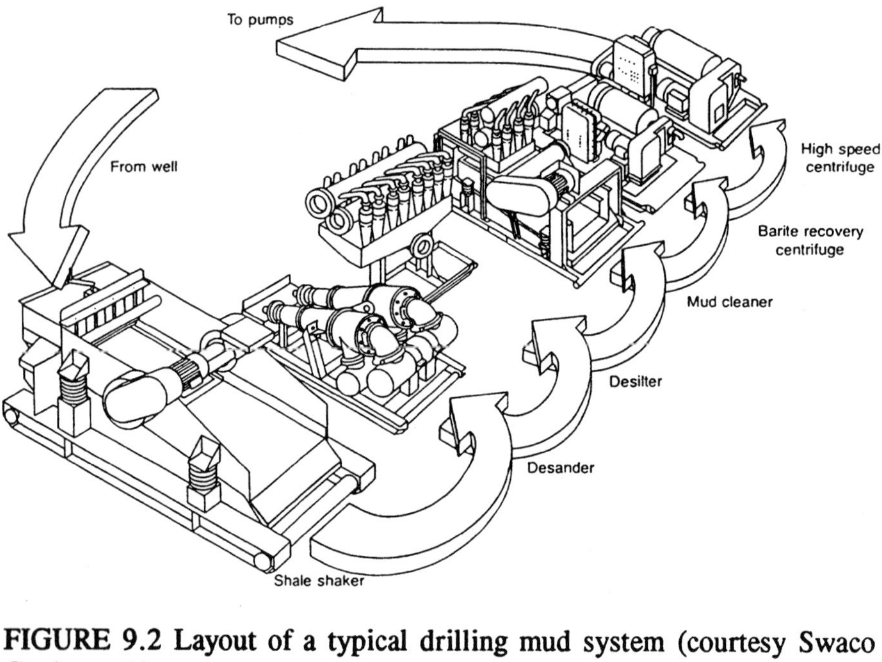

FIGURE 9.2 Layout of a typical drilling mud system (courtesy Swaco Geolograph).

solids is routed from the well to a shale shaker, where the large solids are removed. As the fluid is routed to each successive piece of equipment, progressively finer solids are removed (desander, desilter, mud cleaner, and centrifuge). Desanders and desilters work on a hydrocyclone principle shown in Figure 9.3, whereas centrifuges work on mechanically induced centrifugal forces as shown in Figure 9.4. Proper design and installation of this equipment are critical to the operation of the system (Ormsby, 1973). Solids that remain in the fluid beyond this point cannot be removed mechanically. If they become a problem, the fluid must be discarded and replaced or treated with chemical flocculants.

Other problems are present within the current system that impede performance (e.g., by limiting the speed at which a hole is drilled) or increase the cost of materials handling, including the following:

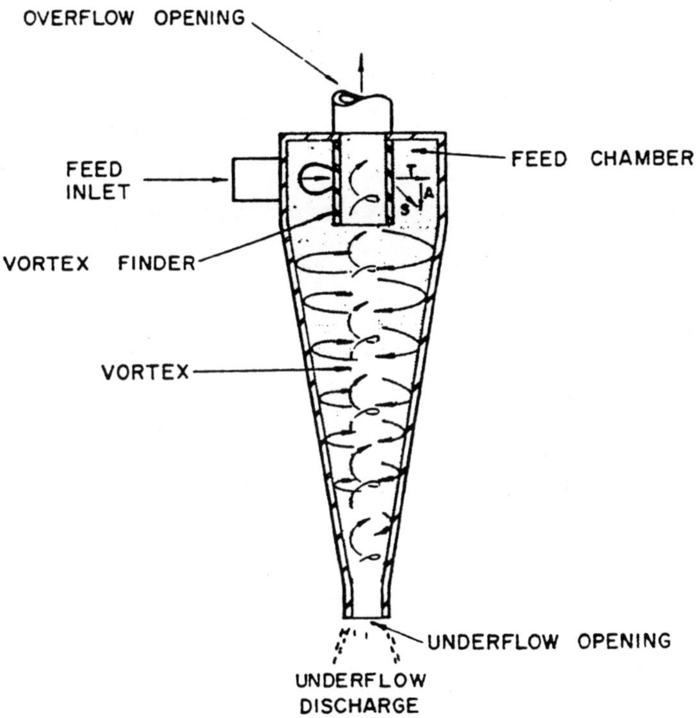

FIGURE 9.3 Schematic cross section of a hydrocyclone (Moore, 1974).

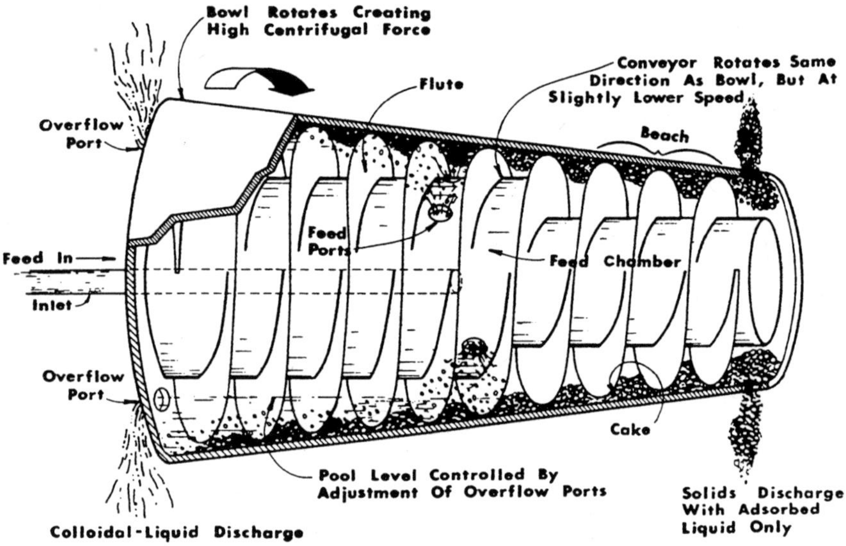

FIGURE 9.4 Schematic of a decanting centrifuge (Moore, 1974).

1. *Lost circulation:* When circulation of the fluid is lost, the drilled cuttings are not removed. They can accumulate and halt the drilling operation.

2. *Incompatibility:* The drilled cuttings may not be compatible with the other functions of circulating fluid, such as density for well control or hole stabilization.

3. *Environmental disposal and contamination:* The fluid can be environmentally hazardous (oil), or the drilled cuttings themselves may be hazardous (e.g., they may contain heavy metals). Disposal of the fluid or cuttings then becomes difficult and expensive, and mixing of the fluids and cuttings contaminates both systems.

4. *Information pathway blocked by choice of fluid*: The fluid system must aid in obtaining information from the well while drilling (such as directional and logging information) and during non-drilling time (such as characterization of the formations by electrical logging, stress state

measurements, or fluid analyses). Using air and aerated fluids interferes with this information transfer (particularly measurement while drilling), and the use of oil or brine interferes with other measurements (such as electrical logging).

5. *Hostile environment:* Hostile environments (high temperatures or pressures) interfere with the effective removal of material.

Even though these problems are present, current solids control systems are capable of separating the cuttings that come to the surface through mechanical or chemical (flocculation) separation. Current fluid circulation methods are capable of moving the cuttings to the surface. The problems are almost always related to other competing functions that the fluid system must fulfill, which may limit the allowable velocity, viscosity, density, or particle size.

Status of the Field for Mining and Tunneling

In mining and tunneling, the volumes of materials to be moved are greater than those in drilled wells by three or four orders of magnitude. In addition, individual pieces of material—often called muck—typically have dimensions up to several inches on a side. Removal becomes a problem of mass movement of very coarse materials at rates up to a few hundred cubic yards per hour. These materials are handled by three systems:

1. a system that removes materials from the advancing face, which may include buckets, augers, conveyors, or pipes;
2. a system that moves cuttings laterally from the face to a shaft or other access point, which may include trains, trucks, conveyors, earth movers, or pipes; and
3. a system that moves cuttings vertically at the shaft, which may include lifting muck cars, buckets, skips, conveyors, or pipes.

Although the practice is not common, muck can be brought to the surface in a pipeline entrained in fluid. In these cases, the problems encountered are similar to those found in drilled wells. The much greater volume of coarse materials from mining or tunneling is usually carried to the surface as loose or broken material. At the surface, the materials are stored in stockpiles or bins until they can be removed and used or disposed of in an environmentally acceptable manner. Problems with the present removal systems include the following:

1. *High-volume, reliable removal:* With the volumes of muck removed in mining and tunneling being hundreds of cubic yards per hour, current systems are marginal in their ability to keep up with a rapidly advancing tunnel boring machine or mining operation, especially in small tunnels. Experience at the Superconducting Super Collider, for example, indicates that state-of-the-art tunnel boring machines may produce muck faster than the best available mucking system can remove it. Any breakdown in the material removal system causes immediate cessation of underground operations.

2. *Compatibility with other functions:* Workers and material must use the same work site entrances (shafts) that are used to bring excavated materials to the surface. Safe and efficient shared facilities must be provided.

3. *Environmental disposal:* Mucking or surface workings are often noisy and dirty operations handling large volumes of materials. At a remote mine, this may be of less concern, but at projects in populated areas (such as urban infrastructure projects), these operations can be disruptive to the environment and to other surface activities. Any new system must be compatible with environmental requirements while handling large volumes of materials.

Priorities for R&D

For well drilling, mining, and tunneling operations, materials handling cannot be considered in isolation from the drilling system. For well drilling, cuttings are usually small and mixed with a transporting fluid, whereas for tunneling and mining, the muck usually contains large pieces and requires mass movement techniques. Advanced materials handling systems must be designed to:

- remove muck or cuttings at the face;
- transport the muck or cuttings from the face to the surface; and

- provide for recovery, decontamination, and disposal.

All of these functions must be compatible with other drilling or tunneling activities.

The development of improved materials handling systems requires advancements on several fronts. These are, broadly, the following:

1. *Improved drilling fluids:* In the short term, research is needed to develop environmentally "friendly" drilling fluids, such as foam and air-mud fluids, with material removal properties equivalent to those of oil-base muds, which are the current state of the art for rapid removal of cuttings but present difficult disposal problems. In the long term, these efforts should lead to the development of drilling fluids that can be varied through the density range between gas and rock and can be adjusted easily as needed to remove cuttings while balancing the pore pressure to prevent loss of circulation. Such fluids must be capable of transmitting acoustical energy for data transmission. They should also have sufficient viscosity at low shear rates to carry large-size cuttings, while having a low viscosity at high shear rates to prevent adverse effects on penetration rate.

2. *Improved muck removal systems:* Current muck removal systems cannot efficiently handle more than about 400 ft of tunnel advance per day. In the short term, these systems should be improved to handle tunnel advance rates of at least 500 ft/day. Long-term (decadal) efforts should anticipate significant improvements in tunneling capabilities that will result from the development of advanced tunneling technologies; such systems should be capable of handling tunnel advance rates up to 2,000 ft/day in ideal materials.

3. *Contaminated materials handling:* Research on contaminated materials handling should focus on the development of automated systems for reducing the volume of contaminated waste through either separation or decontamination techniques. Long-term research efforts should be undertaken to minimize waste handling through in situ disposal of muck or cuttings.

References

Bourgoyne, A. T., Jr., Millheim, K. K., Chenevert, M. E., and Young, P. S., Jr., 1986, Applied Drilling Engineering: SPE Textbook Series, v. 2, Richardson, Tex., SPE, 502 pp.

Moore, P. L., 1974, Drilling Practices Manual: Houston, Tex., PennWell Publishing.

Ormsby, G., 1973, Proper rigging boosts efficiency of solids-removing equipment: Drilling, SPE Reprint Series, no. 22, Richardson, Tex., SPE, p. 147-153.

Rodgers, W. F., 1963, Composition and Properties of Oil Well Drilling Fluids: Houston, Tex., PennWell Publishing.

Walker, R. E., and Mayes, T. M., 1975, Design of muds for carrying capacity: Drilling, SPE Reprint Series, no. 22, Richardson, Tex., SPE, p. 139-147.

10
CONCLUSIONS

Technology to drill holes and to excavate tunnels and openings in rock is vital for the economic, environmental, and scientific well-being of the United States. Drilling is a key technology in several applications of strategic or societal importance, including energy and mineral production, environmental protection, and infrastructure development. During this century, U.S. technology has dominated the worldwide drilling industry and much of the excavation and comminution industries. In the committee's view, this U.S. dominance is likely to erode without continued technological advances.

Although incremental improvements in the component processes in the present state of the art can continue to make drilling more productive, it is the basic conclusion of this committee that revolutionary advances are within reach through the introduction and concerted development of *smart drilling systems*. A smart drilling system is one that is capable of sensing and adapting to conditions around and ahead of the drill bit to reach desired targets. This system may be guided from the surface, or it may be self-guided, utilizing a remote guidance system that modifies the trajectory of the drill when the parameters measured by the sensing system deviate from expectations.

The smart drilling system does not currently exist, but it is presaged by recent dramatic advancements in directional drilling and measurement-while-drilling technologies. Rapid innovation in microelectronics and other fields of computer science and miniaturization technology holds the prospect for greater improvements—even revolutionary breakthroughs—in these systems.

The development of smart drilling systems has the potential to revolutionize drilling. Research in this area will have a significant impact on drilling success and overall cost reduction. Such "smart" systems are increasingly needed to overcome the drilling challenges posed by small, elusive, easily damaged subsurface targets. This is particularly true in applications where identification of small or difficult-to-predict drilling targets and formation damage are key issues in drilling success.

In the development of smart drilling systems, the required improvements in the sensing elements of the system will have an impact on other processes such as rock breaking, debris removal, and borehole stabilization. Revolutionary advances in these fundamental processes might be possible as information about the subsurface environment becomes available in real time.

The development of a smart drilling system requires concerted technological advances in several areas, which include the following:

1. *Development of precise connections between measurable properties and local drilling resistance:* Such connections must be established based on the existing understanding of the connection between rock constitution and comminution mechanisms. Many physical and microstructural properties such as porosity, elastic properties, and wave attenuation can be measured locally. These must be associated more precisely with factors that govern the drilling resistance of rock through directed mechanistic studies and modeling to develop automated response characteristics of a smart system.

2. *Development of sensors for the smart drilling system:* The smart drilling system requires sensors that are capable of detecting and measuring the following:

- *Conditions at the drill bit:* Sensors are needed for in situ measurement of pressure (including pore pressure), temperature, permeability, mineralogic and chemical composition of the rock and heterogeneities, borehole fluid composition (at the part-per-million level for environmental applications), stress state, and rock strength.
- *Conditions ahead of the drill bit:* Sensors are needed that measure rock properties (e.g., porosity, elastic properties, wave attenuation) ahead of the drill bit to adjust drilling parameters, such as the weight on the drill bit and the rotary speed, and to avoid potential problems (e.g., blowouts or loss of circulation) while drilling.

- *Spatial position of the drill bit:* Sensors are needed that are capable of detecting the position of the drill bit in space as required, in order to steer the bit around undesirable zones and reach desired targets.

3. *Development of control systems* for accurate positioning and steering of the drill bit and for automatically adjusting drilling parameters (e.g., load and torque on the drill string, flow rates of drilling muds and fluids), according to local conditions, is necessary to optimize rock breakage and rock removal. This will require precise information at the bit-rock interface of the rock breaking mechanism, rock strength, pressure, temperature, and stress state.

4. *Development of improved methods for steering the drill bit:* A number of mechanical methods for steering are currently available, but large turning radii often preclude their application in a smart drilling system. To enhance steering capabilities, R&D is needed to develop downhole motors, flexible drill strings, and guidance techniques for smart drilling systems.

5. *Continuous monitoring of the state of the entire drilling unit,* including wear of tools, state of other mechanical components, flow of coolant, and the like, is required to anticipate the occurrence of possible discontinuities.

6. *Development of improved telemetry methods for transmitting real-time borehole data to the surface:* The use of advanced sensors for real-time downhole measurements will require significant improvements in data telemetry. Such telemetry is essential for monitoring the smart system from the surface. At present, the most advanced telemetry systems utilize mud-pulse technology and are capable of transmitting data at only a few bits per second. Rates on the order of kilobits per second or higher will be required for advanced smart drilling systems. Telemetry is a rate-limiting step in present drilling systems, and it will become more so as smart drilling systems are developed.

7. *Development of means for continuous and instantaneous support of the rock around the borehole:* The support provided by the rock itself should be used, where possible, in lieu of casing the hole as a separate operation.

Although the principal thrust of the proposed R&D program should be on smart systems, the program should also facilitate incremental improvements in all consequential aspects of present drilling technology. This should result in more immediate attainment of greater efficiencies and cost savings, and will lend needed justification to continued long-term support for the program of smart system development. Such additional R&D should focus on the following problems:

- novel drilling technology, with a focus on the physics of rock removal to reduce energy requirements for drilling;
- improved cutter materials and bearings;
- improved bits for drilling in heterogeneous materials;
- development of environmentally benign drilling fluids; and
- development of durable, compact, high-power downhole motors for directional and extended reach drilling.

This R&D program should be a national effort. Both the public and the private sector should benefit; resources and guidance for the program should be shared, where appropriate. This program should have the following characteristics:

- Integration of industry, university, and government perspectives should be achieved.
- Federal support should serve primarily as a catalyst, with industry providing both technological and financial support. The percentage of R&D support from the federal government and industry could be project specific. The actual R&D should be done by the best-qualified institution whether in the private sector, universities, or government laboratories.
- Finally, a long-term commitment is needed to accomplish the objectives of the program.

The program should be structured with shared research objectives among the federal and industrial partners. Support of projects should be based on a peer-review process and assessment of how the results would contribute to overall program goals. Competition for research funds should be open to industry, national laboratories, and universities.

Attainment of the proposed enhanced drilling capabilities through both short-term and long-term R&D requires a long-range administra-

tive structure that combines the discipline, mission orientation, and flexibility needed to nurture the required scientific and technological innovations. Although the committee discussed a number of possible administrative structures, it ultimately concluded that recommendations in this area were outside its task and expertise.

APPENDIX A

STATEMENT OF TASK

National Research Council
Commission on Engineering and Technical Systems
Commission on Geosciences, Environment, and Resources

Advanced Drilling Technologies

The committee is expected to evaluate the technical and scientific feasibility of advanced drilling and related technologies. The committee will examine concepts for new mechanical and nonmechanical drilling applications including advances in knowledge of the tool-rock interaction, will identify potential opportunities for research, and will make recommendations in a report on the scope and direction needed to realize these opportunities for improved methods for drilling rock. The committee will restrict its recommendations to specific research opportunities and avoid general advocacy of greater expenditures on drilling research.

The committee will consist of 8-10 experts drawn from the engineering and scientific disciplines associated with the topic, including geotechnical engineering, rock mechanics, soil mechanics, material sciences, geology and geophysics, mechanical engineering, and drilling technology.

The committee will meet about four times over the course of a one-year period. Included among the committee meetings will be a workshop to which 20-25 additional experts will be invited. The workshop will be an opportunity for the committee to raise preliminary findings and opinions for discussion among a larger group of experts and to elicit novel or unconventional ideas for penetrating rock. The workshop will not produce a separate report; results of the workshop will feed into the committee's final report.

APPENDIX B

Workshop Attendance

Michael Adams	Federal Highway Administration
Bernard Amadei	University of Colorado
Ali S. Argon	Massachusetts Institute of Technology
John Aslakson	Gas Research Institute
Jamal J. Azar	University of Tulsa
David Biancasino	U.S. Department of Energy
Martin Chenevert	The University of Texas, Austin
Brian Clark	Anadrill/Schlumberger
John M. Cook	Schlumberger Cambridge Research
Neville G.W. Cook	University of California, Berkeley
George A. Cooper	University of California, Berkeley
Kevin D. Crowley	National Research Council
Paul Dauphin	National Science Foundation
Calvin K. Deem	Amoco Production Company
Perle M. Dorr	Meridian Corporation
James Dunn	Sandia National Laboratories
Steven Glaser	National Institute of Standards and Technology
Rick L. Graff	Chevron Services Co.
Sidney Green	TerraTek, Inc.
William J. Gwilliam	U.S. Department of Energy
Michael M. Herron	Schlumberger-Doll Research
Stephen A. Holditch	S.A. Holditch and Associates
Ercill Hunt	Ercill Hunt and Associates
Allan Jelacic	U.S. Department of Energy
Arnis Judzis	B.P. Exploration
Dawn S. Kaback	Savannah River Technology Center
John M. Kemeny	University of Arizona

Billy Joe Livesay	Livesay, Inc.
Stephen E. Laubach	The University of Texas, Austin
Paul Lurie	BP Research Laboratory
William C. Luth	U.S. Department of Energy
David Malone	Anadrill/Schlumberger
Ralph L. Maness	Oryx Energy Company
William C. Maurer	Maurer Engineering, Inc.
Ian McGregor	National Science Foundation
John (Ted) Mock	U.S. Department of Energy
James E. Monsees	Parsons Brinckerhoff, Inc.
André Piché	Noranda Technology Center
Jonathan G. Price	National Research Council
D. Stephen Pye	UNOCAL Corporation
Mark Rankin	Pool Energy Services, Inc.
Richard J. Robbins	Robbins Company
Jean-Claude Roegiers	University of Oklahoma
John C. Rowley	Pajarito Enterprises
David Russ	U.S. Geological Survey
Frank J. Schuh	Drilling Technology, Inc.
Bill Sharp	Colorado School of Mines
Eugene D. Shchukin	Institute of Physical Chemistry RAS
Michael Sheppard	Schlumberger Cambridge Research
Peter H. Smeallie	National Research Council
Jeffrey B. Smith	U.S. Environmental Protection Agency
Steve Souders	U.S. Environmental Protection Agency
Ronald P. Steiger	Exxon Production Research Company
Michael A. Storms	Texas A&M University Research Park
David A. Summers	University of Missouri-Rolla
Thomas M. Usselman	National Research Council
J. Kim Vandiver	Massachusetts Institute of Technology

Ray Wallace	U.S. Geological Survey
Teng-Fong Wong	State University of New York, Stony Brook
Ching H. Yew	The University of Texas, Austin
Albert Yost	U.S. Department of Energy
Mario Zamora	M-I Drilling Fluids Co.
Mark D. Zoback	Stanford University